GROW GREAT VEGETABLES IN *New York*

GROW GREAT VEGETABLES IN *New York*

Marie Iannotti

TIMBER PRESS
Portland, Oregon

Frontispiece: Few things are more rewarding than harvesting homegrown vegetables from one's own garden.

Photography and illustration credits on p. 235.
Published in 2019 by Timber Press, Inc.

The Haseltine Building
133 S.W. Second Avenue, Suite 450
Portland, Oregon 97204-3527
timberpress.com

Printed in China
Cover design by Amy Sly and Adrianna Sutton
Text design by Sarah Crumb

ISBN 978-1-60469-882-4
A catalog record for this book is available from the Library of Congress.

CONTENTS

ACKNOWLEDGMENTS

Learning to garden is an enthralling, never-ending process made all the more enjoyable by the many, many people who generously share the knowledge they have gleaned from both their triumphs and their disappointments. I thank them all. I appreciate every tip, every piece of advice, every seed or seedling, and especially every taste I've so graciously been given.

I am especially grateful to the folks I've met and worked with at Cooperative Extensions. What a wonderful organization filled with the most enthusiastic, helpful people. Whenever I have a question, they have an answer.

I can only imagine what goes on after I write the words. I am in awe of and so grateful to Julie Talbot, Cobi Lawson, and the team at Timber Press, whose persistent optimism make imposing tasks seem like a cakewalk—not unlike gardening.

And, as always, huge thanks to my incredible husband, Michael, for never questioning my sanity as deadlines approach, in print or in the garden.

Becoming a seasoned vegetable gardener takes a lot of digging in the soil and digging for answers. I hope this book helps you with both.

PREFACE

I can't look through a seed catalog without choosing enough varieties to plant a football field–sized garden, with an appetite for more. Just thinking about vegetable gardening makes me hungry. Few things in this world can compete with biting into a freshly picked fruit or vegetable. The scents, the vibrancy, and the anticipation of that eruption of flavor make growing food an all-sensory delight.

Very few edible plants can't be grown in New York, especially if you are willing to push the seasonal envelope. Leafy greens, earthy root crops, luscious berries, and hearty winter squash are all possible. This climate provides gardeners a warm, sunny summer and plenty of chill days for those exacting plants like rhubarb that need a rest between seasons (kind of like us gardeners). We take a brief pause to celebrate the holidays, and then we reach for our seed catalogs and the gardening season is back underway.

Vegetable gardening allows us to be part of the seasons and their changes. Although some people mark spring by the whims of a mercurial groundhog, there is no denying that spring has begun when we see the first green shoots of spinach, asparagus, or rhubarb. It's not summer until we can bite into a beefy, glowing tomato, and just when the garden is overflowing with abundance in early fall, the shortening days remind us that it is time to slow down. The New York vegetable garden may go under cover for the winter—under mulch, under plastic, under snow, or underground—but the process never ends; it just keeps re-creating itself in a most comforting, and often frustrating, way.

Our part of the world is the perfect place to enjoy the change of seasons, and each season brings its own reward. The information offered here will help you make sure you do not miss out on any of the gardening enjoyments the state has to offer, whether it is filling your winter home with sprouting greens and luscious fruits or the succession of harvests from the first spring thaw through the closing curtain of frost in the fall. New York may be thought of in some places as urban and industrial, but it is also home to some of the best farmers' markets, locavore restaurants, and resilient gardeners who can turn any abandoned lot or alleyway into a feast for the soul.

Having four true seasons offers the down time needed to plan and prepare the garden year, and getting the most from a vegetable garden does require a little advanced planning. It's all laid out for you in this book, including a year-round schedule to show what you can prepare, plant, and harvest each season. Whatever your level of gardening experience, you can jump in at any time and get up to speed. Read and digest it all at once, and keep it handy to guide you month by month.

Viewing your garden throughout the year is the best way to learn the intimate nuances of what each plant wants. That knowledge will serve you well, because no two years are the same. It is the rare spring when we can enjoy the gardening tradition of planting our peas on St. Patrick's Day, but it is not unheard of. What I most hope you will glean from this book is the cycle of the garden and the role you play in keeping it moving forward. New York offers a climate for savoring everything from arugula to zucchini; think about what you love to eat, and get ready to take it to another level.

My Garden

Get Started

Whether you're maximizing space in a single raised bed or transforming your backyard, a vegetable garden gives you lots to look forward to.

WELCOME TO GARDENING IN *New York*

Most folks from out of state hear "New York" and think of Manhattan, however there is a surprising array of geographic regions upstate. New York has its share of mountainous regions and bodies of water, all of which contribute to its long, cold, snowy winters and hazy, hot, humid summers. The soil tends to depend on what the retreating glaciers left behind—it could be the black dirt of Orange County or the rocky clay a few counties north. Each area has its growing challenges, but agriculture is still one of the state's top industries and the lure of fresh produce has more and more people growing their own, even in New York City. The growing season may be unpredictable, but it is also rewarding.

◀ Beets grow in a Woodstock organic garden.

Climate Zones

Asking a gardener what zone he or she gardens in is an invitation for commiseration or envy. Plant hardiness zones tell us the length of an area's growing season—when we can expect the first and last frosts of the seasons, which is crucial information for any gardener. Not all growing seasons are equal, however. Knowing the arc of your area's particular growing season will help you extend the time in which you can grow and harvest vegetables. You'll know when it's safe to get started with frost-tolerant, leafy greens; when you can expect them to fizzle out and need replacing with plants that like it hot; and how long you can expect to enjoy every harvest.

In the mid-1900s, the United States Department of Agriculture (USDA) mapped out the United States, Canada, and Mexico into 11 zones based on the lowest annual minimum temperature. Each zone represents a 10°F difference in that low temperature. This ranking made it possible for the people who grow plants to label them according to zone, so gardeners would know which plants should survive winter in their area. New York residents garden in USDA plant hardiness zones 3, 4, 5, 6, and 7.

New hardiness maps were released in early 2012, also using data collected over a 30-year period. Older maps were grouped by the lowest temperature ever recorded, but the newer maps average the data, making it more representative of actual fluctuating climates. The maps now include two new zones at the warm end of the chart. Although these new zones don't apply to New York, most areas here did move up a half zone warmer, a factor attributed to better mapping and weather tracking.

Although hardiness zones provide valuable information for a gardener, they are still just guidelines. Every year is different, and many microclimates fall within zones—and sometimes within your own yard. Exposure to winds and sunlight can play games with the climate in your garden. Think about that shady spot in the far corner where the last of the snow stubbornly refuses to melt, or the south-facing border of your house where the first daffodils put in an appearance each spring, weeks before they show color in other areas of your yard. You can even create your own garden microclimates by putting up walls and wind breaks. An enclosed courtyard or walled garden can retain enough heat to bump the spot into a different zone. Even the radiant heat offered by your home's foundation will affect the hardiness of the plants growing near it.

Obviously, you have more to consider than your hardiness zone and the length of your growing season, but having a reasonable idea of your area's first and last frost dates will be a big help in planning your garden. The accompanying table provides frost dates for several areas in New York.

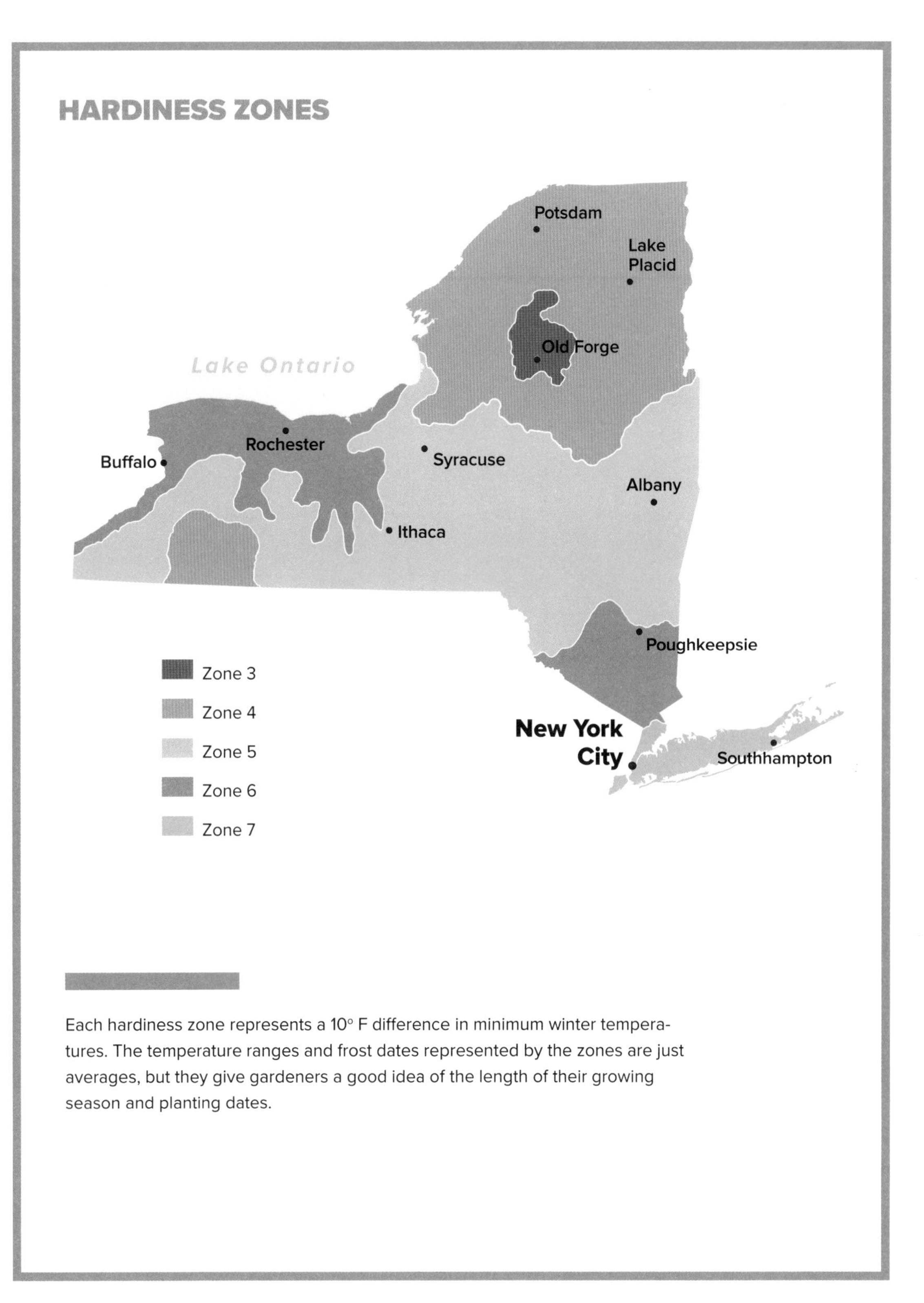

Each hardiness zone represents a 10° F difference in minimum winter temperatures. The temperature ranges and frost dates represented by the zones are just averages, but they give gardeners a good idea of the length of their growing season and planting dates.

AVERAGE FROST DATES IN NEW YORK

CITY OR AREA	FIRST FROST	LAST FROST
Albany	1–10 Oct	21–31 May
Buffalo	11–20 Oct	21–31 May
Ithaca	1–10 Oct	1–10 June
Lake Placid	11–20 Sept	21–30 June
New York	1–20 Nov	21–30 April
Old Forge	1–10 Sept	1–10 July
Potsdam	21–30 Sept	11–20 June
Poughkeepsie	1–10 Oct	21–31 May
Rochester	11–20 Oct	21–31 May
Southampton	21–31 Oct	11–20 May
Syracuse	11–20 Oct	21–31 May

Growing Season

The length of the growing season is the time between your area's average last spring frost date and average first fall frost date. The shortest growing seasons in New York are in the high upstate altitudes, while New York City and Long Island in the southeastern corner are more temperate and have longer growing seasons.

Zones 3 and 4. In upstate New York, the combination of being farther north and higher in elevation translates into cool, damp summers and cold, windy winters. In this short growing season, vegetables must mature quickly.

Average minimum winter temperatures:
Zone 3: –40 to –30°F
Zone 4: –30 to –20°F

Average days over 86°F:
Zone 3: 8 to 14
Zone 4: 15 to 30

Zones 5, 6, and 7. These growing zones encompass most of the southern portion of New York.

Average minimum winter temperatures:
Zone 5: –20 to –10°F
Zone 6: –10 to 0°F
Zone 7: 0 to 10°F

Average days over 86°F:
Zone 5: 31 to 45
Zone 6: 46 to 60
Zone 7: 61 to 90

A community garden in Coney Island flourishes.

NEW YORK REGIONAL GROWING PROFILES

New York City metropolitan area (zone 7). The metropolitan area is considered to have a humid, subtropical climate. The extensive paving and lack of air circulation caused by rows of tall buildings makes the city excessively hot in summer and offers limited space to do any gardening, although hardy spirits have been reclaiming vacant lots and working wonders. More gardening goes on in the outer boroughs which enjoy a long growing season, with plenty of rain and sunshine. However, soils can be contaminated from development and pollution and should be tested before growing edibles.

Long Island (zone 7). Long Island is the only area of the state that is a true zone 7. The western end tends to be warmer because of its proximity to New York City. The eastern end of Long Island gets cool breezes off the water and has summer conditions that are very similar to England's temperate climate. Winter weather can be extreme, but does not last long. The unique, sandy soil is unusually good for growing vegetables.

Hudson Valley and Catskill region (zones 5 and 6). Weather conditions in this area vary not just from north to south, but also the farther you get from the river and the higher the elevation. However, in all areas the weather is unpredictable. Some winters are snowy, others barely register. Summers are always humid, but rainfall is uncertain. Soils vary

Pole beans are harvested in a Woodstock garden.

greatly, from rocky clay to sand, but sustain many regional small farms, apple orchards, and backyard gardens.

Capital/Mohawk (zones 4 and 5). This area is close enough to New England to experience many of the storms coming in from the Atlantic. Winters fluctuate between bitter, snowy cold and lingering thaws. Summers are typical of the Northeast: hazy, hot, and humid. Thankfully, the soil left behind by the glaciers is moderately rich and perfect for growing all kinds of plants. Vegetables do quite well here.

St. Lawrence Seaway (zones 4 and 5). Sandwiched between the Adirondack Mountains to the east and the highlands of the Finger Lake Region, this area is a remote and rural part of the state. Dotted with dairy farms and apple orchards, the seaway retains a natural quality not seen in many areas of New York. The area closer to Lake Ontario has a microclimate that is great for growing vegetables and fruit. It is slow to warm up in the spring, but also slow to cool down in the fall, creating a leisurely growing season.

Adirondacks (zones 3, 4 and 5). Climate and topography team up to make this a challenging area for any type of gardening. The growing season can end in the blink of an eye, with first frosts as early as August. Soil tends to be infertile, acidic, and rocky, more so the higher you go. New York residents voted to keep the Adirondack region "forever wild" and there is still a lot of heavily forested area, but land is being cleared. While the mild summers are not ideal for vegetable gardening, shorter season varieties will do fine here.

Central New York (zone 5). Expect considerable rainfall and cloudy days during summer, then a wallop of snow in the winter, accompanied by strong winds. Conditions here can be harsh, to say the least. However, summer temperatures are generally moderate and pleasant. Soil quality varies widely. Previously forested areas are rich and well draining, while the brownish red soils colored by shale and minerals can make tilling a garden a workout.

Finger Lakes (zones 5 and 6). No part of the state was as uniquely carved out by glaciers as the Finger Lakes. To the north, the landscape is dotted with drumlins, hills shaped like the overturned bowl of a spoon. To the south, the elevation rises and the rolling drumlins give way to steep inclines. This is where you will find the famed vineyards. While the soil can be lean or rocky, the growing season is long and the weather is very hospitable for vegetable gardening. Then there are the deep, open gorges, inhospitable to any type of gardening.

Genesee Valley (zones 5 and 6). The receding glaciers left behind extremely fertile soil in the Genesee River valley. Great soil and a long growing season make the area well suited for growing vegetables and fruit. Although spring tends to come late, the growing season stretches well into fall. Summers are humid but sunny, with enough sporadic rainfall to keep plants healthy and growing long enough to mature.

Niagara Frontier (zone 6). Long, cold, snowy winters start with lake-effect snows in late fall. Both spring and fall can be very unpredictable with sudden frosts. Despite those conditions, the area is renowned for home gardens. Offsetting all of the snow and frost are pleasant, sunny summers that tend to be cooler and less humid than much of the state and offer ample opportunity to grow a wide variety of vegetables.

Chautauqua–Allegheny (zones 5 and 6). From the Allegheny foothills to the shores of Lake Erie, this area is dominated by forest and rich soil. With Lake Erie nearby, lake-effect snow is commonplace, followed by enough snow to keep the reservoirs full. Spring and fall are cool seasons and even during the summer, the temperature rarely gets warmer than the 80s. The soil is quite fertile and the growing season is long enough to enjoy a steady harvest.

GARDENING 101

Unfortunately, plants are not "plug-and-play." They don't thrive if you ignore them, and they require a lot of follow-up, which is why some people love gardening and others avoid it. As with anything worth doing for the long haul, it helps to start with a good foundation. For plants, that would be the essential three elements of sun, soil, and water. Of those three, sun is the one element you will have the least control over, and it's something you should wisely consider before you plan or plant.

◂ Growing veggies means getting to know your soil.

Sun

Vegetables vary in their need for sun. Those grown for their leaves and roots, such as spinach, lettuce, radishes, and beets, do not demand a full day of sunshine. They can be happy with the respite of some afternoon shade. Fruits and fruiting vegetables, such as tomatoes, squash, eggplants, and peppers, need lots of sun to be productive.

Garden books and nurseries usually specify a plant's requirement for sunlight in three particular ways: full sun, partial sun or partial shade, and full shade. Here's a quick breakdown of what these terms mean.

Full sun At least 6 full hours of sun each day. Heat-loving vegetables such as eggplants, okra, and melons want as much sun as they can get. If you've spent long periods of time out in the sun, you know that 6 hours in the morning can be cooler than 6 hours in the afternoon. It's all considered full sun, but keep in mind that if your plants are being exposed to sunlight during the hottest parts of the day, they will also need more water.

Partial sun or partial shade There is no explicit definition here and little agreement about how to make the distinction. In general, this encompasses 3 to 6 hours of sunlight each day, with partial sun at the high end and partial shade at the low end. Plants requiring partial sun are more tolerant of the accompanying heat. Plants requiring partial shade tend to prefer some protection from the afternoon sun. Dappled sunlight that filters through tree canopies and trellises can be treated as partial shade.

Full shade When it comes to plants, full shade means less than 3 hours of sunlight each day and dappled sunlight the rest of the day. It does not mean no sunlight at all.

Before you choose a location for your vegetable garden, watch the movement of the sun across your yard to get a clue of where the sunlight falls at what time of day: for example, a bright, sunny spot at 6 a.m. might be shaded by your neighbor's fence by 10 a.m. Many gadgets are available to measure the amount of sunlight your yard receives—from disks filled with light-sensitive dyes to electronic meters. The low-tech way to calculate your garden's sunlight exposure is to make an hourly chart. Start around sunrise and check at the beginning and end of each hour, noting whether the area receives full or partial sunlight and when it is totally shaded. Then add up the total hours of sunshine. By tracking the sun throughout the day, you'll notice the shading effect of nearby buildings, fences, and other tall structures.

Another circumstance to consider is the growth rate of nearby trees. The maple you planted last year will become a shade tree within 10 years. (It will also send its roots into your well-watered and fertilized vegetable garden.) When it comes to deciding whether to move an established vegetable garden or cut down a mature tree, the tree usually wins.

One final complication for gauging sunlight involves where and when you are gardening. Not all sunlight is equal. The days don't just go from short spring days, to long, lingering summer days, and back to short again. As the sun rotates, it gets closer to the Earth during the summer, and the North Pole is tilted toward the sun. Northern latitudes have the most hours of sunlight in the summer. So, for example, a garden in Bangor, Maine, will receive 15:36 hours of sunlight on the summer solstice, the longest day of the year. Its counterpart in Harrisburg, Pennsylvania, will receive 15:03 hours of sunlight on the same day. This is just the opposite at the winter solstice, when southern gardens see more sunshine than those in the north. Because plants respond to day length, the hours of sunlight become really important, especially if you are trying to get plants to mature in the fall or keep things going with protection in the winter. To take the best advantage of all the available sunlight, position your garden and cold frames to make the most of the sun.

GROWING DAYS MEAN HEAT

You've chosen the perfect location, and the sun is out and shining down upon you, but your plants are languishing rather than thriving. Why? The missing ingredient could be heat. Some vegetables, such as tomatoes and peppers, are actually semitropical plants that need plenty of heat to ripen. So even if your packet of seeds says that fruit will be ready to harvest in 75 days, during a cool season, you may have to wait a little longer.

Cherry tomatoes need sun and warmth to ripen.

Many plants do not start actively growing until temperatures reliably remain at 50°F or warmer. The agriculture industry measures heat as growing degree days (GDDs). In a nutshell, they take the average of the daily highest and lowest temperatures and subtract it from 50, the baseline minimum temperature for growth. Warm-season vegetables such as tomatoes and peppers need a certain amount of these heat units to ripen and develop their sugars and flavor. If you have a long string of cool days, warm-season plants are going to bide their time. On the other hand, vegetables that prefer cooler temperatures such as spinach and radishes will throw in the towel when temperatures rise too high.

Once summer gets going, New York is assured of having more than enough GDDs to ripen tomatoes. But this concept will become more important to you as you try to stretch your growing season by planning late fall harvests and gardening under cover throughout the winter.

Soil

Professional gardeners are taught early that garden soil is never called dirt. Soil is the foundation and future of your garden plants. Dirt is what gets on your clothes. Don't treat your soil like dirt.

It is almost impossible to have healthy plants without healthy soil. Don't be fooled by its inert appearance; there's a party going on down there. Soil is an ecosystem that provides a home for all sorts of insects and microorganisms that, in turn, provide air and nutrients for your plants to uptake. These organisms feed on the organic matter in the soil, such as compost, manure, and decaying plants and insects, and turn it into humus, which improves the soil's texture, retains moisture, and is rich in the many nutrients plants need.

Most soil is made up of organic matter along with crumbling rocks, which are in plentiful supply in the Northeast and New York. Rock provides the inorganic material in soils, and the predominant type of rock in your area will factor into what type of soil you have. For example, limestone creates soil high in calcium with a slightly alkaline pH. Shale often results in a clay soil with an acidic pH. The minerals in the rock also play a role. Rock with lots of quartz can produce the more irregularly shaped grains of soil found in sand. Even topography and climate get involved—for example, smaller grains of soil wash off slopes, excessive rain leaches nutrients, and arid, baked soil is no place for earthworms.

Given the diversity of the topography in our part of the world, it is no wonder that no single type of soil predominates. In my small area, the glaciers left one town with a surplus of shale, the next with sand, and then carried the rich, black sediment into the next county.

Building good soil is an ongoing task and not a quick one. Soil is constantly in transition and can become depleted by rain, chemicals, and even the plants growing in it. If you live in farm country, your soil has probably been amended over the years. However, if you live in some type of housing development, you can bet that the topsoil was removed and sold off during initial construction. A soil test will tell you everything you need to know about your soil. The good news is that any type of soil can be amended and made suitable for gardening.

SOIL TYPES

Good soil is a combination of texture, structure, and fertility.

Texture refers to the size of the soil particles.

Sandy soil is comprised of large, irregularly shaped particles. This makes it feel coarse in your hand and prevents it from compacting easily, so water and nutrients percolate right through it. Sandy soils are usually low in fertility.

Clay is made up of flat microscopic particles that pack together, preventing water and air from circulating. Add heat, and clay soils become bricklike. On the plus side, most clay soils are highly fertile.

Loam is considered a balanced soil, and this is what you are striving for in your garden. It is rich in organic matter, has a healthy structure, and attains the elusive moist but well-draining quality preferred by most plants.

Structure refers to how well the soil holds together. A good soil structure is crumbly and allows plant roots to branch out and water to

Assessing your soil means not being afraid to get your hands dirty.

drain but not run off. Grab a handful of damp soil and form it into a ball. If the ball totally falls apart with a slight tap, the soil is sandy. If it stays intact, you have clay. You want something in between.

Fertility is the most dynamic aspect of soil. The nutrients in your soil are the measure of its fertility, and nutrients are continually being used up by plants. Luckily, the means by which you replace nutrients are the same as those you use to improve your soil's texture and structure—adding organic matter. Organic matter refers to decomposing plant and animal materials, and in gardens here that generally is compost, composted manure, and green manures (cover crops). Adding organic matter gives substance to sandy soil and lightens clay. It feeds the microorganisms and insects in the soil and even makes plants' roots more permeable so they can take up more nutrients and water.

SOIL PH

Nothing you do for your soil will make an appreciable difference if the pH is too high or too low to sustain your plants. Soil pH matters because plants can access nutrients from the soil only if it is within a specific pH range; if the soil pH is too far out of that range, the nutrients will sit idly in the soil and will never become available to the plants.

The abbreviation pH is shorthand for potential hydrogen, which is a measurement of hydrogen ions in a solution. Don't worry if this means nothing to you; just know that the pH scale, with a range of 0 to 14, is used to determine the relative acidity or alkalinity of a substance—in our case, soil. A pH of 7.0 is considered neutral. Anything below a 7.0 is considered acidic, or sour, and everything above 7.0 is alkaline, or sweet. You might think it's no big deal if your soil's pH is only 1 point above 7.0, but this is a logarithmic scale, so each number represents a power of ten. That means that a pH of 8.0 is ten times more alkaline than a pH of 7.0.

DO-IT-YOURSELF SOIL TEST

You can get a sense of the quality of your soil just by looking at it and observing how well plants grow in it, but periodically testing for nutrient levels will give you a better idea of its fertility and what amendments might be needed. You can start with this simple home test to determine your soil's texture. A predominately loamy soil—a balance of sand, clay, and silt—is a gardener's dream; it will retain water and provide a base of organic matter.

WHAT YOU'LL NEED:

- Small scoop or trowel
- Clean jar with straight sides and a tight-fitting lid
- Water
- 1 Tbsp. powdered dishwashing detergent
- Soil sample

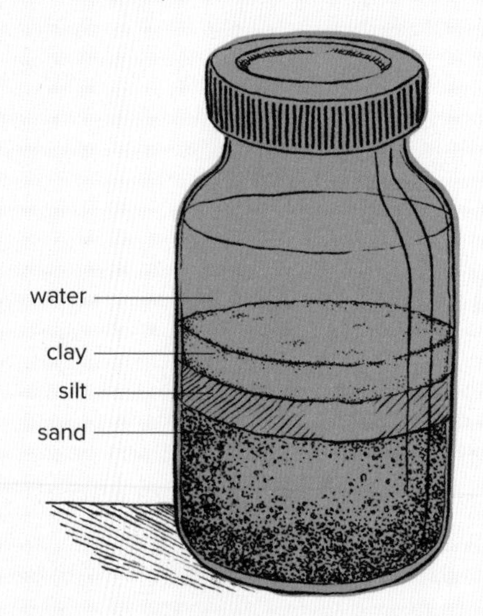

▲ Know the percentage of sand, silt, and clay in your soil, and you can estimate its water- and nutrient-holding ability and determine what needs improving.

STEPS:

1. Collect small soil samples from three or four sections of your garden and mix them together.
2. Sift the sample to remove stones and debris.
3. Fill the jar about halfway with the soil.
4. Add the detergent to the jar; this will prevent the soil particles from clumping together.
5. Fill the jar about three-quarters full with water and tighten the lid.
6. Shake vigorously, making sure that all clumps are broken, nothing is stuck to the bottom, and everything is well mixed.
7. Set the jar on a level surface and check back periodically over the next couple of days to see how everything has settled.
8. Sand is heaviest and will settle to the bottom of the jar.
9. Silt will form a layer on top of the sand and will be a bit darker than sand.
10. The tiny clay particles can take days to settle. This layer will be lighter than the silt.
11. Any stray organic matter should float to the top.

As gardeners, we need to be concerned with pH, because different plants prefer different soil pH levels; in general, plants are happy in soils between pH 6.2 and 7.0. I wish I could tell you that all New York soils fall into this range, but most tend to be on the acidic side (below 7.0) of the scale. Luckily, you can change that.

But before you can adjust your soil's pH, you need to know what its pH actually is. Many home-testing kits are available, and some are more accurate than others. You can also bring a sample into your local cooperative extension office and have them test it. Armed with that knowledge, you can start making adjustments. You can improve an acidic soil by applying lime. For alkaline soils, you can add some type of sulfur (usually elemental sulfur), aluminum sulfate, or iron sulfate. These are not instant fixes, however. The best time to amend the soil to improve its pH level is in the fall; this allows the lime or sulfur time to work its way into the soil in time for spring planting. These are also not one-time fixes. You will need to monitor your soil pH continually and make adjustments as needed.

FEEDING PLANTS

Plants are living things, which means that just like you, they have to eat. They do pull some nutrition from the soil we have taken great pains to build, but these nutrients will become depleted if they are not replenished. You can add plant food, or fertilizer, to amend the soil and provide the plants with the nutrition they need so that they can do the same for us.

Good things come in threes, and when it comes to plant food, the big three are nitrogen (N), phosphorus (P), and potassium (K). Every bag of fertilizer comes labeled with the amount by weight of each of these key elements, always in the same order: N-P-K. If you see 5-1-1, for example, you know immediately that the fertilizer in the bag contains 5 percent nitrogen, 1 percent phosphorus, and 1 percent potassium.

A balanced fertilizer, in which the percentages of all the elements are fairly equal, such as 5-4-4, is recommended for general use. However, you'll find specialty fertilizers with higher ratios of a particular element that target a particular plant's needs. For instance, leafy vegetables need a fertilizer high in nitrogen. Phosphorus encourages more flowering and fruiting, an obvious benefit to many garden edibles. These elements are all vital and work together, but each has its own strengths.

Nitrogen assists in the development of chlorophyll and protein and is necessary for healthy green leaves. Without enough nitrogen, leaves will slowly turn yellow and the plant's growth will be stunted. Too much nitrogen will encourage a lot of leafy growth and little flowering and fruiting. Nitrogen is the most rapidly depleted element in the soil.

Phosphorus contributes to many plant functions, including root growth, overall health, and the setting of flower buds and the resulting fruits. Phosphorus depletes slowly, but it is accessible to plants only when the soil is warm and when the soil pH is within a 5.0 to 7.0 range. Working it into the soil near the plant's root zone helps the plant access it more easily. Phosphorus deficiency can manifest as excessively dark green leaves, purplish leaves or stems, and fruits that drop before they mature.

Potassium, sometimes referred to as potash, is essential for a plant's overall healthy functioning—from strong growth, to disease resistance, to the quality and flavor of the fruits and vegetables. It tends to stay put in the soil, so you do not need to apply it often. But when you do, apply it near the roots and work it into the soil. Some signs of potassium deficiency are yellowing along the leaf veins and leaves that roll up or crinkle.

Assisting this triumvirate of nutrients are many micronutrients required for healthy plant growth, including calcium, magnesium, sulfur, iron, and zinc. Most of these micronutrients are supplied by fertilizer mixes or compost, so let's talk about choosing a fertilizer.

THE SCOOP ON FERTILIZER

Every year, new choices in plant food appear on the store shelves, and most are moving toward organic. Using organic, slow-release fertilizer in the vegetable garden is a win-win situation. You don't have to worry about what you're putting into your future food, and the organic materials actually feed the soil in addition to the plants. Most synthetic fertilizers contain salts that wash away in the soil and can harm that ecosystem you worked so hard to develop. They do nothing to build up healthy soil, so they must be continually reapplied. Applying an inorganic fertilizer is similar to trying to survive on vitamin supplements instead of eating food. Organic fertilizers enrich and amend the soil, releasing their nutrients slowly so that they are available when the plant needs them.

You can find complete slow-release organic fertilizers to work into your garden beds a couple of times a year. In addition, water-soluble organic fertilizers can be used through a drip irrigation system or as a quick feed from a watering can. Fish emulsion and manure tea, for example, can be applied to provide a shot of nitrogen for plants that need it.

Because soil varies greatly from region to region, many gardeners prefer to have a complete soil nutrient level test done to learn exactly what their soil needs and then add just those amendments. (You can get your soil tested at a lab or via your local cooperative extension office.) Once you know exactly what you need, you can find organic amendments at a good garden center. Most are by-products reclaimed from the farming and fishing industries. Keep in mind that when it comes to feeding plants, more is not better. Follow the recommended doses.

Alfalfa meal (2-1-2) usually comes in pellet form. Alfalfa is a nitrogen-fixing legume—which means it holds onto nitrogen in nodules on the roots. In addition to adding some nitrogen to the soil, alfalfa meal provides potassium and many trace minerals and growth stimulants.

Blood meal (12-0-0) is a slaughterhouse by-product. It is one of the highest nitrogen supplements you will find, but use it with caution, because applying too much at one time can burn plants. Don't let it come in contact with leaves. Blood meal also makes a decent deterrent for herbivores such as rabbits and groundhogs, but it can attract dogs.

Bone meal (3-15-0) is ground animal and fish bones. A great source of phosphorus and calcium, it also adds some nitrogen and micronutrients. Like blood meal, it can attract dogs and also rodents. Some studies suggest that the phosphorus in bone meal is available to plants only in soils with a pH lower than 7.0.

Cottonseed meal (6-0.4-1.5) is another good source of nitrogen. Be sure the product is labeled organic, because pesticide residue stays in the cotton seed.

Feather meal (12-0-0) is another slaughterhouse by-product. It is very high in insoluble nitrogen, which is released into the soil slowly, so not all of it washes away in the spring rainy season.

Fish by-products are rich sources of nitrogen. Fish emulsion (5-2-2) is a balanced, water-soluble food with the misfortune of smelling rather foul. The odor fades, but it can attract fish lovers such as bears to your garden. Fish meal (10-6-2), made from dried, ground fish, is not as foul smelling but is slower to work.

Greensand (0-0-3) is mined from prehistoric marine deposits in New Jersey. Containing 3 percent potassium and many micronutrients, it breaks down very slowly and makes an excellent long-term soil conditioner.

Mulching helps conserve water and keep roots cool.

Kelp/seaweed (1-0-2) amendments made from dried, ground seaweed are valuable for their micronutrient content. Liquid kelp is often mixed with fish emulsion to create a complete fertilizer.

Oyster shell flour (NPK is negligible) is a mix of coarse and finely ground oyster shells. It contains 96 percent calcium carbonate and a healthy dose of micronutrients. The nutrients are released slowly and do all kinds of wonderful things for your plants and your soil, including assisting with the uptake of fertilizer, aiding cell growth, regulating pH levels, and improving soil texture. It also makes a beneficial amendment to compost heaps and worm bins.

Rock phosphate (0-3-0) is also known as colloidal phosphate. It is made by encasing clay particles with insoluble phosphate and provides a slow release of 2 to 3 percent phosphate throughout the growing season. It also adds micronutrients to the soil. Another form, called hard rock phosphate, has a higher percentage of phosphorous (0-33-0), but more of it is released into the soil and it isn't as beneficial to the home gardener.

Soybean meal (7-2-1) is an excellent source of nitrogen and phosphorus and can usually be purchased where livestock feed is sold. Soybean meal is a nice alternative if you do not or cannot use fish products.

MULCHING

I find mulching a tedious chore, but I'm always glad I did it, because it makes such a difference. Mulching with a 2- to 4-inch layer of material around your plants helps prevent

START A COMPOST PILE

The composting process in a compost pile helps speed up the decomposition of organic matter. Build a pile of all your garden debris and watch it turn into a sweet-smelling, crumbly, free enhancement for your soil. You have to weed, prune, and clean out your garden anyway, so why waste all that free fertilizer?

You can actively compost by maintaining the right balance of ingredients and moisture so that the pile heats up and decomposes quickly, or take the passive approach of simply piling up your garden trimmings and allowing them to decompose naturally, over time. With an active compost pile, you'll need to add alternating layers of brown, carbon-rich materials such as dried leaves, and green, nitrogen-rich debris such as kitchen scraps and grass trimmings (2 parts brown to 1 part green). Don't be concerned about having to keep the right balance of brown to green; sooner or later, it all has to be mixed together anyway, so if your preferred method is to dump in whatever is handy at the moment, that will work, too—but it may take a little longer to become finished compost. Here are the basics of composting to get you started.

CREATE AN ENCLOSURE

This is not absolutely necessary, but containing the material will help it heat up and decompose faster. The enclosure can be constructed from wood or it can be as simple as a circle of wire fencing, such as chicken wire. You can even compost in a garbage can if you drill some holes for air to get through and water to drain. Whatever you choose, it should be large enough to accommodate a pile of at least 3-by-3 feet so the pile will heat up sufficiently—but don't be tempted to make it much taller than 4 feet or it will be too heavy for airflow, not to mention turning.

CHOOSE A LOCATION

Choose a warm location that gets plenty of air circulation but not too much direct sunlight.

WHAT TO ADD

- **Brown material:** Leaves, straw, shredded paper, and small twigs.
- **Green material:** Grass clippings, garden debris, vegetable kitchen scraps, coffee grounds, and eggshells.
- **Air:** Oxygen will speed the decomposition and help prevent the buildup of odors.
- **Water:** Keep the pile slightly moist. It should hold about as much moisture as a wrung-out sponge. Compost that is excessively wet will have an unpleasant odor and will leach nutrients.

CREATE THE PILE

1. To create layers, start with about 4 to 6 inches of brown material, such as leaves.
2. Top with 2 or 3 inches of green material.
3. Repeat steps 1 and 2.
4. When your pile reaches about 3 feet tall, stop adding new material.
5. Use a garden fork to stab and twist through the compost to mix it up, and flip the outer portions of the pile into the center. Mixing the pile, or turning it over, will add oxygen and drain excess water, speeding up the composting process. How often you choose to do this is up to you.
6. If your pile develops an unpleasant odor, it has become anaerobic (oxygen deprived) and needs more oxygen and/or less water. An unpleasant odor is a sure sign that the

pile needs to be turned. On the flip side, you may need to add moisture to the pile during hot, dry weather, especially if your compost is exposed to the sun. Covering it with black plastic helps hold in moisture and prevents rain from leaching off nutrients, but getting the right amount of moisture without making the pile anaerobic is a balancing act.

7. As the pile decomposes, it will start to heat up. Microbes such as bacteria and fungi oxidize and break down the high-carbon brown matter, generating substantial heat as a by-product. The center of a pile with a good balance of ingredients can reach temperatures of 140 to 160°F. As the heat increases, it creates an environment for more efficient microbes to take over and speed up decomposition. Turning the pile will make sure that the microbes are getting sufficient oxygen and that the whole pile heats up. Compost piles that are not actively maintained probably won't get hot, but they'll still decompose eventually. The benefits of a hot pile go beyond speed, however. A hot compost pile will also kill off insect eggs, plant diseases, and most weed seeds.

FINISHED COMPOST

When the pile cools off, your compost is ready to use. This can take a couple of months to a year. It should be dark brown in color and should smell fresh and feel crumbly in your hand. If a few chunks of branches remain, you can sift the compost before applying it to your garden. Use finished compost to amend the soil in the spring and to side dress around plants while they are growing. You'll never have enough compost, so start making a new pile as soon as you can.

COMPOSTING RULES OF THUMB

- Layers should be about 2 parts brown to 1 part green.
- Cover smelly kitchen scraps with a layer of brown or another type of green to deter pests.
- Do not compost meat, fish, dairy, fats, or human or pet waste.
- Avoid diseased plants, weeds going to seed, and anything that was treated with an herbicide.

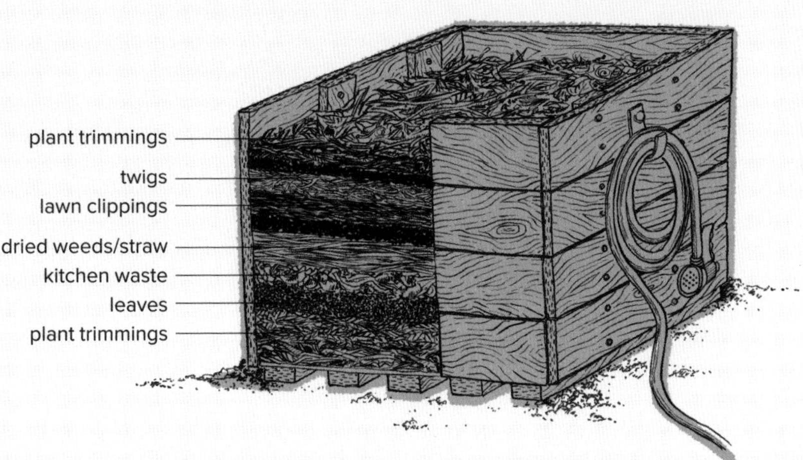

◀ Alternating layers of green and brown materials in your compost pile speeds up the composting process.

weed seeds from germinating, conserves water, and keeps the soil and plant roots cool. Mulch also keeps soil from splashing up onto your plants when it rains. One hour spent spreading mulch will save you many hours of weeding and watering later in the season.

Inorganic and organic mulches are readily and inexpensively available throughout the state. Inorganic mulches range from plastic sheeting to stone. Even better, organic mulches will slowly decompose and feed the soil and are made of almost any available plant material, including straw, pine needles, shredded leaves, compost, and wood chips. Because you will be starting from scratch each spring, straw makes an excellent vegetable garden mulch that can be easily moved aside for planting and seeding. It lasts the entire season, and you can turn it into the soil to decompose. It also attracts spiders that hide out and eat insect pests.

Compost makes a great mulch and adds a great deal to the soil by improving its water-holding ability; it also helps maintain a balanced ecosystem and makes for healthier plants. But it can be expensive to buy if you haven't made enough of your own to cover your entire garden. And weed seeds will eventually germinate in it.

Water

Plants need water even more than we do. They use it to access nutrients in the soil, and they need lots of it to plump up their fruits and roots. Plants in drought situations become stressed, making them susceptible to diseases and pests. Badly stressed plants go into survival mode by dropping their flowers and fruits, stopping growth, and focusing all their reserves on keeping themselves alive.

Although rain or snowfall is generally plentiful in New York from fall through spring, summer droughts are not uncommon. Many New York gardeners rely on home wells for all their water needs, and they must be judicious in how they use that water. During long dry spells, the water level in wells can get precariously low, with no relief in sight. Soaking the garden may not be a water use priority. But even folks who pay for city water are getting it from a well: all water comes from the ground and is a limited resource, so there is no reason to waste it.

We can let our lawns go dormant and choose drought-tolerant ornamental plants, but all vegetables and fruits need lots of water. Luckily, there are plenty of sensible options to help keep your plants watered without racking up a brain-exploding utility bill or depleting your well.

RAIN BARRELS

Rain barrels connected to your home's gutter downspouts can collect and store rainwater. Barrels are fairly easy to connect and can fill up surprisingly fast. They come in assorted capacities, with 60 gallons being the most commonly available. Make sure your barrel has a screen on the top to filter debris and keep breeding mosquitoes out. You'll need a way to access the water, and a built-in spigot with a hose attachment will make this much easier.

WATERING CANS

Although they are not an efficient means of watering an entire vegetable garden, watering cans are sometimes indispensable, such as when you need to water a couple of containers or distribute water mixed with fertilizer. I recommend you have a couple cans, because

A watering can should have a rose that allows a gentle flow of water.

it's easier to walk while balancing one in each hand. I prefer a watering can with a long handle that extends from the front of the top opening to the bottom of the can; I can slide my hand along the handle, making it easier to pour. A removable rose is also handy; a rose with small holes creates a fine stream of water for watering young seedlings, but you can remove it when you want the water to pour out in a stream.

GARDEN HOSES

Garden hoses are a necessary evil; they are heavy and difficult to roll up, and if you are not careful, a hose will flatten every plant in its path as you drag it around. But hoses are the most efficient and flexible way to get water to your garden. Do yourself a favor and invest in a high quality, kink-resistant hose. And if you'll be leaving it outdoors all year, get a "no freeze" hose. Good quality rubber is heavier than vinyl but it lasts much longer. You might also invest in several quick-connect fittings; they snap together and apart in seconds. You'll also need a nozzle with at least two settings: a fine mist or shower for watering, and a strong spray for obliterating pests.

SPRINKLERS

Sprinklers are quick and easy water distributors. I leave a sprinkler in the garden and snap on the hose when needed. If an extended dry spell is expected, you can use a timer to distribute water to the sprinkler during the hottest parts of the day. All sorts of sprinklers are available, but I prefer those that offer various watering patterns. When choosing a sprinkler, keep in mind that overhead watering is not ideal: it wets the leaves, which can make them susceptible to fungus, plus a lot of water is lost to evaporation and poor aim.

DRIP IRRIGATION AND SOAKER HOSES

Drip irrigation and soaker hoses distribute water slowly and can be placed so that water is delivered directly at the plants' roots. Drip irrigation kits and parts are available in garden centers and hardware stores, and most are simple to pop together to create any layout to suit your needs. If you set up your garden in the same pattern year after year, you can configure your system, keep it intact, and lay it out quickly in the spring. Because the ground freezes throughout much of New York, you'll need to set it up in the spring and pack it away in the fall, which is a lot of effort, but once you've set up an irrigation system you can use a timer and forget about it. Soaker hoses—perforated hoses that ooze water—can be moved around easily, and this provides more flexibility than a drip system.

Runner Beans

GARDEN PLANNING

Complaining about the weather is winter sport for Northeasterners, but our long, cold winters are really a mixed blessing. Although they keep us indoors, we have plenty of time to plan our gardens, and it's easy to get carried away. If this is your first vegetable garden, you can be realistic or be overwhelmed. If it's not your first garden, think back to last season's tomato plants that flopped over their stakes and the cabbage your kids refused to eat. Do you want a repeat of that? Instead of fantasizing, grab a cup of tea and a seat by the window and ask yourself, How much time do I have to spend in the garden? How much space can I spare? What vegetables do I like to eat? Your answers lay the foundation for your garden plan.

◀ A little planning ensures you'll be happy with your harvest.

Space and Time

The best location for a vegetable garden is a sunny spot that is free of tree roots and other obstructions and preferably near a source of water. Somewhere close to the kitchen is ideal but not always possible. Traditionally, vegetable gardens have been tucked out of sight in the backyard, but vegetable gardens in front yards are becoming more popular these days, and front yards often receive more sunlight than backyards. Remember, however, that it may take extra work to maintain a front-yard vegetable garden to keep it looking good for curb appeal. You'll probably need to install some type of fencing if your yard doesn't already have it, because whether you garden in the back or front, animals will find your cucumbers wherever you plant them.

After you settle on the perfect spot for your garden, be honest with yourself about how much time you can devote to its maintenance. In addition to the initial prepping of the soil and planting, you will need to commit to staking, weeding, reseeding, harvesting, and monitoring continually. Vegetables do not wait until you have time to pick them. Will you have time to harvest before dinner? Are you willing to spend Saturday mornings weeding and feeding, even in 90-percent humidity? If your answer is a resounding yes, take the plunge. If work, carpool, and committees make your schedule unpredictable, don't give up; just start small, with a few tomatoes, some lettuce, and a carefree pumpkin or squash.

Deciding What to Grow

Making choices about what to grow is probably the most important of your garden planning tasks, because it will determine how much space and time will be required. Everyone tries to plant more than their space can hold. That is a gardening given, and it really doesn't get any better with experience. However, you can make your decisions easier by considering a few things.

PLANT WHAT YOU LIKE TO EAT

This should be common sense, but seed catalogs are seductive. Looking at a photo of backlit Swiss chard might make it seem tempting to grow, even if you've never really cared for cooking greens. Heirloom eggplants look very inviting, but who wants to eat eggplant every week during August? Think about how you already eat, not an idealized version of what the back-to-the-earth gardener eats. If your kids love carrots, leave space to plant more carrots every couple of weeks. If you are really interested in salad vegetables, start with tomatoes, lettuce, cucumbers, and radishes. You can always add more.

PLANT WHAT YOU CANNOT GET FRESH LOCALLY

Sweet corn is grown up and down the East Coast, and the start of sweet corn season is a regionally celebrated event. The idea of growing your own sweet corn is alluring, but corn takes up a lot of space in the garden, and caring for it consumes a good deal of time and resources. You may be better off forgoing the corn, because farmers' markets and stands will be overflowing with it.

Asparagus, another farmers' market staple, is a perennial crop. To grow asparagus, you'll need to create a space where it can grow undisturbed for 20 years or more. That might mean a trade-off. Are you willing to sacrifice a section of your garden to be able to snap off spears minutes before dinner, or would it be more practical to stop at the farm stand for asparagus and leave space for more tomatoes?

On the other hand, matchstick French green beans are rarely grown commercially and taste their snappy best when eaten fresh from the garden. Home gardeners can grow thousands of vegetable varieties that will never make it to the produce aisle. Planting something new and different every year is a great way to expand your culinary horizon as well as your gardening chops.

You may decide to buy corn at local markets because of the garden space it requires.

PLANT THINGS THAT GROW WELL IN YOUR AREA

Most of New York gets enough warm growing days to sustain just about anything. However, some crops, such as okra, sweet potatoes, melons, and eggplant, need a long growing season. Most melons require several months of hot weather before they even start setting fruits. If you live in zone 5 or lower, you can try a short-season variety, or you can leave the melon growing to the professionals.

In addition to climate considerations, pests and diseases are perennial problems. Colorado potato beetles seem to have potato radar. Powdery mildew will seek out your squash and cucumber plants. Multiple leaf-spot disease spores will lie dormant in the soil, waiting for the perfect conditions to infect your tomatoes. Look for vegetable varieties that are labeled as resistant to recurring problems in your area. Choosing resistant varieties does not guarantee immunity, but it does give you an edge when weather conditions are right for troubles to take hold. There are plenty of choices, and I have included a good assortment of varieties in the "Edibles A to Z" section.

DIRECT SOWING VS. TRANSPLANTING

PLANTS USUALLY DIRECT SOWN

- Beans
- Beets
- Carrots
- Corn
- Cucumbers
- Lettuce
- Melons
- Parsnips
- Peas
- Radishes
- Squash (including pumpkins)
- Turnips

PLANTS THAT TRANSPLANT WELL

- Asian greens
- Basil
- Broccoli
- Brussels sprouts
- Cabbage
- Cauliflower
- Celery
- Chives
- Eggplant
- Kale
- Kohlrabi
- Leeks
- Mustard
- Okra
- Onions
- Parsley
- Peppers
- Swiss chard
- Tomatoes

Seeds vs. Seedlings

Another part of your garden planning process is deciding whether you will plant seeds or purchase seedlings. The New York gardening year is too short to sow seeds directly in the garden for plants that require 4 to 5 months to mature. We would never see a harvest of tomatoes, peppers, onions, Brussels sprouts, or parsley if we waited until May to sow seed. Gardeners solve this problem by starting their seeds indoors under lights or by purchasing seedlings.

You can start sowing seed indoors in February, and this is a nice way to get your hands back in some soil, even if the soil outside is buried under a foot of snow. Starting plants from seed allows you to choose exactly the varieties you want to grow and greatly expands your options. Even so, not everyone enjoys the months-long process of growing plants indoors, and not everyone has the room for it. Thankfully, nurseries and farmers' markets spill over with seedlings throughout the spring. You can roll through, scoop up some seedlings, and no one will be the wiser as the plants thrive in your garden. If you start with seedlings, look for a reputable nursery or seller that offers locally grown plants. Many pick-your-own farms also offer vegetable seedlings in the spring; they sell the same plants they have chosen for their gardens, based on what grows well in the area.

Unfortunately, not all vegetables transplant well from pot to garden soil. Some will have to be started from seed sown directly in the garden. Plants grown for their roots, such as carrots and radishes, and those with long taproots, such as corn, do not like to be disturbed while they are growing, so direct sowing is best. Peas, beans, squash, and other

SPACE GUIDELINES FOR VEGETABLES PER SQUARE FOOT

ONE PLANT	FOUR PLANTS	NINE PLANTS	SIXTEEN PLANTS
Broccoli	Corn	Beans	Arugula
Cabbage	Herbs	Beets	Carrots
Cauliflower	Potatoes	Peas	Lettuce
Cucumber	Strawberries	Spinach	Onions
Eggplant	Swiss chard		Radishes
Pepper			
Tomato*			

**More space may be required for large varieties*

quick-growing vegetables do not benefit from starting their seeds early, either. Plant these seeds in the garden, and they'll catch up to transplanted seedlings in very little time.

Garden Design

You know that edible plants need rich, healthy soil with lots of organic matter worked in and at least 6 hours of sun each day. And you have found the perfect spot to create your garden. Now what? Although aesthetics play a major role in ornamental garden design, an edible garden requires a few different design considerations. Vegetable gardens can be attractively designed, but they are always in a state of flux, so pairing plants based on what looks good together is not particularly important. Your two main concerns should be making sure your plants have the room and sunlight they need to grow and providing adequate access for harvesting them.

Thinking about adequate space for your plants takes some finagling. Looking at the tiny seed and seedling, you might lose sight of how large the plant will be when it reaches maturity. It helps to have some idea of plant spacing before you venture out into the garden. You can do this in your head or, even better, on paper. Planning is nothing fancy—just use graph paper and divide your garden plot into square-foot increments based on the printed lines. Large plants such as tomatoes need an entire square foot per plant. Plants that keep their basic footprint but fan out, such as chard and basil, can be planted four to a square. Plants started from scattering seeds can be grown a little closer and possibly thinned out as they grow.

The accompanying table provides a space guideline for vegetables using one, four, nine, and sixteen plants per square foot.

Raised beds offer a number of advantages for vegetable gardeners.

Planting Options

Gardeners never have enough space for all the plants they grow, and squeezing everything in requires some layout logistics. A handful of planting options let you make efficient use of whatever space you have, while still offering your plants enough light and air and giving you enough room to get in there and harvest.

ROWS OR FURROWS

Growing vegetables in straight rows allows easy access for tending, staking, and harvesting, and you can weed with a hoe instead of bending over to work between plants. By planting in rows, you can reconfigure your garden each season and customize it according to what you want to plant. And sometimes it just makes sense to plant in rows, such as when you need a deep trench for planting leeks so you can easily backfill as the leeks grow. The only downside to planting in rows is that a lot of space remains unplanted to provide access paths between each row.

BLOCKS OR WIDE ROWS

Planting in blocks can solve the unplanted space issue. Rather than planting a long, single row, you can consolidate the plants into a wider area and sow seeds closer together, either by scattering them or creating multiple rows within the block. Either way, you won't need space for excess paths. Continual harvesting prevents the block from becoming overcrowded, and the plants can continue growing without taking up too much space. Rows can be as wide as you like, but avoid making them so large that you can't reach into them to weed and harvest. Blocks and wide rows are especially good for cut-and-come-again crops such as salad and cooking greens. The downside is that plants can become crowded, limiting air and sunlight and leaving them damp and susceptible to fungal diseases.

HILLS AND MOUNDS

In the garden, hills are not mounds of soil; they are a circle of seeds. Happily, the vegetables that need to be planted in hills also like being planted in mounds, so we don't have to sweat the definitions. Squash, cucumbers, and melons need warm, dry soil, and mounds provide just that. They also need lots of flowers for good pollination; planting a circle of 3 to 5 seeds in each mound should produce enough flowers for a good yield.

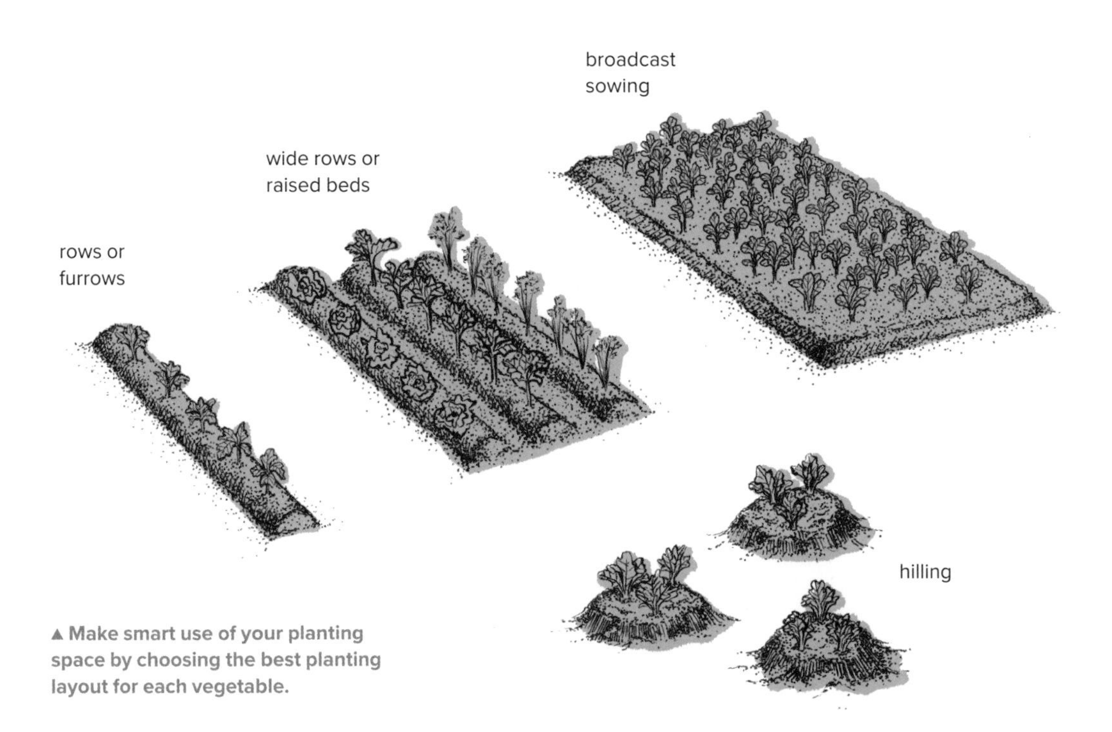

▲ **Make smart use of your planting space by choosing the best planting layout for each vegetable.**

RAISED BEDS

Raised beds are wide rows that are elevated from the ground by several inches to several feet. They can be as simple as soil lifted from the path and piled onto a bed, or they can have some type of retaining walls, such as lumber. Raised beds are generally just wide enough to allow you to reach the center of the bed from any side—usually no wider than 4 feet.

Raised beds have a number of advantages, especially for New York gardeners. Even a 6- to 8-inch depth will help the soil dry out and warm faster in the spring and will improve drainage all season. Because you will not be walking on the beds, the soil will not compact, and no tilling is required. You can amend just the soil within the bed, and there will be less nutrient run-off. And if you make your beds high enough, you can sit on the wall edges while working.

Permanent beds also make planning easier. You know exactly how much space you have and how plants will be laid out—in rows or big blocks of the same vegetable, or mix things up within each bed. You can, for example, place a pole bean teepee in the center of the bed, with a block of beets on one side, a block of lettuce on another, and a row of pepper plants occupying the remaining space.

CONTAINERS

A row of potted herbs on the windowsill may be a common sight, but vegetables outside in containers are becoming equally familiar. Planting in containers can be a solution for a lot of garden dilemmas. If you have poor soil, or no soil, grab some pots and start a garden on your balcony, porch, or driveway. You can locate containers to make the most of the sun. Another big bonus: your fruits and vegetables

RAISED BED MATERIALS

Raised beds can be permanent structures or temporary fixes, such as a mound of soil, although mounds have a tendency to flatten out. You don't, however, want to use anything that might contaminate the soil in which your vegetables are growing.

Wood and resin lumber. Although any type of wood can be used, red cedar and redwood are top choices for garden structures. Wooden bed structures can be held together using screws, nuts and bolts, rods of rebar, or easy slide-in corners marketed specifically for raised beds. Resin lumber, a recycled plastic product often used for decking, can also be used. **Pros:** Easy to assemble. Rot resistant. Lasts for years. **Cons:** Can be expensive. Real wood will eventually need replacing.

Natural stone. If you live in a rocky area and are constantly digging up large stones in the garden, you can put them to good use as the walls lining a raised bed. **Pros:** Durable. No maintenance. Beds can be created in any shape. **Cons:** Heavy to work with and takes considerable effort to build.

Blocks and bricks. Recycled bricks can be repurposed as sturdy walls. Many types of prefabricated blocks are available for building walls, and the interlocking types are especially easy to install. **Pros:** Beds can be created in any shape. Good heat retention. If buried below soil level, can prevent tunneling animals. **Cons:** Heavy to work with.

Poured concrete. If you have identified a location for a raised bed and are committed to it indefinitely, poured concrete is a sturdy and durable option. **Pros:** Little or no maintenance. Retains heat and prevents tunneling animals. **Cons:** Initial installation requires time and effort. Cannot be moved.

Straw bales. One of the quickest, easiest ways to build a raised bed is with stacked bales of straw or hay. **Pros:** Quick and easy to build. Can be reused in the garden as mulch. **Cons:** Will eventually decompose. Large bales can be difficult to reach over and into the garden.

If you decide to create raised beds, keep a few things in mind:

- Consider accessibility and height. Beds for folks who have difficulty bending could be as tall as waist height.
- Leave room for paths and access.
- Several small beds can be easier to maintain and access than one large bed.

▸ **Whether your raised beds are permanent structures or simply mounded soil, make sure you can reach to the center of the bed.**

Spinach and other greens are easily grown in containers.

are up off the ground, where they are less likely to rot or be eaten by assorted creatures.

There is almost nothing you can grow in the ground that cannot be grown in containers. From a tray of salad greens, to a vat of potatoes, to a columnar apple tree, plants will happily perform in confinement if you give them enough space for their roots to spread. Most will need a pot at least 12 inches deep. Larger plants, such as tomatoes and vining squash, will need twice that depth. Five-gallon buckets are a good option, but a half whiskey barrel is my recommendation for large plants like beefsteak tomatoes.

A few words about container size:

- The larger the container, the less often you will need to water it.
- Large containers filled with soil can be very heavy, especially when wet. Either fill your containers in their final resting spot or place them on a plant dolly with sturdy wheels.
- If you are gardening on a deck or rooftop, make sure the structure is capable of bearing the extra load.

Use a good quality potting soil and mix in a granular, organic fertilizer before planting. Granular fertilizers are slowly released and not water soluble, so they do not leach out of the soil every time you water. Depending on the length of your growing season, after one fertilizer application, you may not need to feed your plants again. For heavy-producing vegetables such as tomatoes, however, you might feed them either some water-soluble organic fertilizer when they start setting fruits, or add a little more granular fertilizer to the top inch of soil around midseason.

MAKING THE MOST OF YOUR SPACE

Whatever the size of your garden, you can produce a steady supply of fruit and vegetable favorites by using a few techniques.

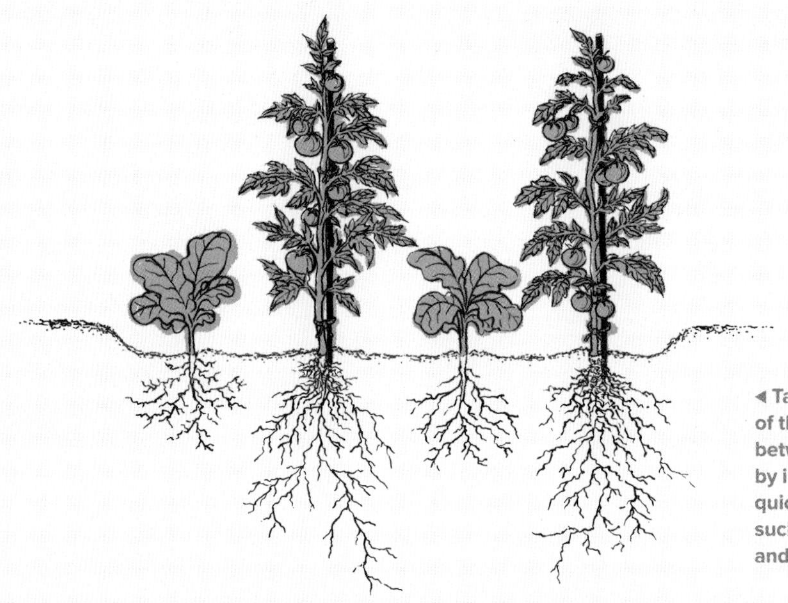

◀ **Take advantage of the open spaces between tall plants by intercropping quick growers such as lettuce and other greens.**

Succession planting. Stagger harvest times by planting a short row of the same vegetable every 2 or 3 weeks. Fast-growing vegetables, such as peas, radishes, beets, and bush beans, are easy to reseed; as you finish up harvesting the first planting, a new crop will be ready to eat.

Plant varieties that mature at different times. Extend the season by planting an early season cabbage such as 'Fast Ball' (45 days) along with a midseason variety such as 'Chieftain' (88 days) and maybe a 'Late Flat Dutch' (110 days). This will keep you in coleslaw without overwhelming you.

Intercropping. Planting more than one vegetable in the same space sounds like a recipe for overcrowding, but if you plant something that matures quickly, such as beets, next to something that doesn't get growing until midsummer, such as eggplant, you have created a good partnership. Or plant lettuce, radishes, and other vegetables that prefer some shade and cooler temperatures in peak summer in the shadow of taller corn or tomato plants.

Keep seeds handy. Fill in any vacant spots as early-season vegetables fade.

A community garden brings together like-minded neighbors.

COMMUNITY GARDENS AND OTHER SHARED SPACES

If you don't have the space or site conditions to grow vegetables at home, you still have options. Check with area garden clubs or with local government to find out if community vegetable gardens are available in your area. Some let you rent your own space and other spaces are grown and tended communally, with each gardener getting a share of the harvest.

Many small farms offer community supported agriculture (CSA) projects. You purchase a share in the community area of the farm, and each week you receive a box full of whatever crops are in season. Most of these farms allow you to volunteer to work at the farm and offer a discount in the price of the share for your efforts.

Another option is to join forces with a group such as a church, local school, hospital, or nursing home to plant a garden in unused space. Involving your community is a great way to ensure the garden's success. You could do the same thing on the neighborhood level. Share the chores and the bounty by making one large garden with your neighbors, rather than each of you growing more zucchini than anyone wants to eat.

Keeping Garden Records

Keeping track of what and when you plant, where you plant it, and how well it grew is a great way to learn what works best in your garden. Each year, you can indicate which

CROP ROTATION

Life would be easier if we could lay out our vegetable garden plan once and simply replant the same things each spring. Unfortunately, planting crops in the same spot each year allows problems to multiply. Pest insects will lay their eggs nearby or directly in the soil, ready to hatch and start feeding the minute you put your seedlings out in the spring. The spores from fungal diseases can easily overwinter, even under snow cover, and reinfect your plants in the dampness of spring.

Even if we could control these lurking problems, repetitive planting depletes the soil nutrients in areas where heavy-feeding vegetables such as corn, cucumbers, and broccoli are planted. Alternating what you grow in each bed allows you to build up the soil's fertility and structure.

Crop rotation is an established organic gardening practice and one of the easiest ways to keep your vegetables growing healthy. Many experts suggest rotating plants by family, since pests and diseases often favor all the relatives in a plant family. For example, cabbage worms enjoy dining on cabbage, but they're also fond of broccoli and kale. Early blight is not just a concern for tomatoes and potatoes; it can affect their nightshade cousins, eggplants and peppers, too. But the length of time and distance of separation for this type of rotation can be impractical for small backyard vegetable gardens. Who can find a new spot to grow tomatoes every year for 10 years?

A much easier approach that still offers the benefits of preventing pest build-up and soil depletion is to group your plants according to the part of the plant you eat. This also roughly organizes them by their feeding requirements: leafy vegetables (nitrogen), fruiting plants (phosphorous), root crops (potassium), and legumes such as beans and peas (nitrogen-fixing). Corn is included with the leafy crops because it is a heavy nitrogen user. Grouping plants with similar nutrient needs also simplifies adding fertilizer and amendments.

YEARLY ROTATION PLAN

LEGUME *Adds nitrogen*	LEAF *Uses nitrogen*	FRUIT *Uses phosphorus*	ROOT *Uses potassium*
Beans	Broccoli	Cucumbers	Beets
Peas	Cabbage	Eggplants	Carrots
	Cauliflower	Melons	Onions & leeks
	Herbs	Peppers	Parsnips
	Kale	Squash	Potatoes
	Salad greens	Tomatoes	Radishes
	Spinach		Turnips
	Corn		

vegetables did well and those that did not thrive. This information can help as you plan for the next growing season.

Ideally, your garden journal will list several factors.

- Dates when seed was started, seedlings were hardened off, plants were put in the ground, first flowering began, first fruits appeared, first harvest started, and last harvest ended
- Notes on the vigor of the plants and the quality and quantity of the yield
- Problems encountered, such as insect pests, diseases, slow growth, and animal problems
- Notes on flavor, special needs such as staking or pruning, and things you want to remember if you plant it again, such as starting the seeds earlier or later

I also like to keep notes on the weather, the amount of rainfall, and some phenological factors. (Phenology is the study of when things in nature reoccur each year.) Make note of these natural indicators and you'll have a good guideline for when to plant or when to be on the alert for certain pests and problem. For instance, dandelions tend to bloom when the soil is warm enough to plant potatoes. Squash vine borers usually lay their eggs about the same time that chicory blooms along the roadside. Variable weather means gardeners can't always gauge our planting dates by the calendar. Plants and animals react to day length and average temperatures, and they can provide a more accurate sense of when it is time to get things in the ground—barring events such as a freak May snowstorm. Common sense is still your best gardening tool.

You can keep track of all this information in many different ways, from a spreadsheet to a bag full of plant tags with notes on them. Two easy, organized ways to keep notes are via online programs and with a digital photo journal. Online journal resources are easy to use and many are free. You still have to make the effort to use them—after all, reports are only as good as the information you put into them.

As a time-crunched record keeper, I find it easiest to keep my camera handy. I take lots of photos throughout the growing season. Because digital photos are time-stamped, I can use the photos later to see when tomatoes ripened and when powdery mildew set in. I even photograph the labels for a record of what I planted. I keep each year in its own folder on my computer, subdivided by month. I also keep all my tags and seed packets of new varieties, so I know what I grew each year. It's not a perfect method, but it's one I can keep up with; plus, the more time I spend in my garden, the more I can intuitively know what to expect.

Get Planting

Seeds and seedlings set the stage for a successful vegetable garden.

SHALLOTS
RED SUN
SHARPES
Allotment
North
Shed.
Cut Flowers
Cold Frame
Early Spuds.
'Arran Pilot'
'Jubilee'
Leeks ?
Winter Bandits
Spindle Trees.
Asparagus Beds
Peas. K. Wonder
Broad Beans
Peas
Broad. Beans.
Shallots
'Red Sun'
Onion set
Red Baron.
Courgettes
(old compost Bin)
Carrots -
Cortina.
Carrots
WITKIEM
MANITA
BB
Black plastic
Grown on
new ground.
site of old Compost Bins.

JANUARY

Planning and Plotting

January may not seem like a logical start to the New York gardening year, but there's nothing like a post-holiday bout of cabin fever to send your garden dreams soaring. Although you won't be working in the garden this month, it's a great time to start thinking about what you want to grow and how you are going to squeeze it all in. A good plan and a flexible schedule are indispensable garden tools, and January is the ideal month to start planning. By midwinter, when the seed catalogs start piling up in the mailbox, we all begin to long for the flavors of summer—tomatoes, basil, fresh greens, and melons. Our dreams may be bigger than our gardens, but we've still got a couple of months to firm up our plans.

◀ Winter is the perfect time to plan your dream vegetable garden.

TO DO THIS MONTH

PLAN

Everyone

- Review notes about last year's garden for winners and losers
- Inventory leftover seeds, test for viability, and list what you need to reorder
- Organize catalogs and make your wish list
- Review your wish list and be realistic
- Start ordering seeds

Zones 5, 6, and 7

- Gather and clean seed-starting materials, such as containers, trays, and soil

PREPARE AND MAINTAIN

- Check for animal damage to fruit trees and shrubs
- Check on vegetables in storage
- Check on hoop house vegetables
- Order asparagus

SOW AND PLANT

Everyone

- Start a windowsill garden with micro-greens, lettuce, and herbs

Zones 6 and 7

- **Sow indoors (late in month):** celery, leeks, onions, and thyme

HARVESTING NOW

From storage

- Garlic
- Onions
- Parsnips
- Potatoes
- Squash
- Turnips

From hoop house

- Arugula
- Beets
- Brussels sprouts
- Carrots
- Herbs
- Kale
- Leeks
- Parsnips
- Swiss chard

Zone 3 Zone 4 Zone 5 Zone 6 Zone 7

Brussels sprouts seeds ready for planting.

Seed Catalogs

Seed catalogs are designed to lure you in. They promise trouble-free tomatoes, non-stop spinach, and a tower of strawberries. Go ahead and indulge in the fantasy. Dream big. Make a list of all the things you want to grow, and then go have a delicious lunch. Come back to that list with a full stomach and a more discerning eye, and pare it down to what will fit in your garden space and what you truly want to grow and eat.

A good way to begin to narrow down your list is to look through your leftover seeds. Start with two piles: plants you would like to grow again and those you can pass on this year. Pull out the packets to see what you already have on hand. Check on the quantity of the seeds you want to grow again. Make notes on your dream list about whether you need to order more or cross them off the list. And be careful not to order too many varieties of the same vegetable—you probably don't need four types of pole beans.

If you garden in raised beds, you know exactly how much growing space is available. Even if you change your garden layout every year, the square footage should remain constant enough for you to estimate how many rows or plants you can reasonably fit in. At this point, you can ruthlessly execute the judgments you made about what your family will actually eat, how much, and how often. If you have limited space, I strongly recommend you do without single-serving space hogs such as cabbage and cauliflower and devote space to vegetables that produce regularly over a long period of time, such as lettuce, cucumbers, Swiss chard, and tomatoes. It's OK to try a couple of varieties of each, but limit the total number of plants. You'll find that spaces are continually opening in the garden, and you can reseed quick-growing crops such as lettuce and beets throughout the growing season.

TESTING OLD SEEDS FOR VIABILITY

You probably have leftover seeds from last year (or even the year before), because it seems the packets always have more seeds than you can use. But before you assume you can plant them for this growing season, you'd be wise to check the seeds for viability. Plant seeds deteriorate with age, and if you stored them in less-than-ideal conditions, they will deteriorate faster. Check how well your old seeds germinate before you place a seed order so you'll know whether you really need more.

1. Take ten seeds from the packet.
2. Lightly dampen a paper towel and place the seeds on it, spaced out in a row.
3. Roll up the paper towel, seal it inside a plastic bag, and put it in a warm spot.
4. Refer to the "Seed germination rates and estimated years of viability" table, or look on the seed packet for a germination time, and start checking the seeds whenever germination can be expected. For instance, broccoli seed should germinate in 5 to 10 days.
5. Count how many seeds germinated and multiply by ten. This will give you the percentage viability of the seeds in the packet. If more than 70 percent germinated (7 of 10), your seed should be fine to use. If only 40 to 60 percent germinated, you can use the seed, but sow it thickly and expect some bare spots. A rate below 40 percent is not really worth planting. Time to order new seed.

Creating a Garden Map and Planting Schedule

After you've determined how much space you have and what you intend to plant, you can map out where everything will go. Because you won't be planting everything outdoors at the same time, your map will ensure that space is available for later plantings. It will also help your work move along quickly when you finally get your hands in the soil. Garden mapping is fairly easy if you stick to the crop rotation plan. Group your plants according to type: leafy vegetables, root crops, fruiting plants, and legumes. Then choose a bed or location for each. If you planted something in one spot last year, simply rotate it forward by one section. I will be honest with you and say that most gardeners break this rule. It is the rarest of gardeners who don't squeeze in plants wherever they find room after the season gets moving. But, at least, you can start off with good intentions and try to stick with them as best you can.

The tricky part of garden mapping is planning for succession planting. If you know you'll be planting lettuce, beans, or beets every 2 or 3 weeks, make note of that on your

Make sure past-season seeds are still viable before planting.

map. You can leave some areas unplanted to be filled in later, or you can designate replacements for early harvests. Spring vegetables such as spinach, radishes, and peas will be finishing up and ready to harvest when it's time to start planting beans and more lettuce. You can continue like this right through to the fall, when harvesting the last of your tomatoes and cucumbers will make room for vegetables you can overwinter, such as kale, carrots, and parsnips.

SEED GERMINATION RATES AND ESTIMATED YEARS OF VIABILITY

	APPROXIMATE DAYS TO GERMINATION	VIABILITY OF SEEDS, IN YEARS
Bean, snap	7 to 10	3
Beet	7 to 14	4
Broccoli	5 to 10	5
Cabbage	5 to 10	5
Carrot	12 to 15	3
Cauliflower	5 to 10	5
Corn	7 to 10	2
Cucumber	7 to 10	5
Eggplant	10 to 12	4
Kale	5 to 10	4
Lettuce	7 to 10	5
Melon	5 to 10	5
Onion	10 to 14	1
Parsnip	14 to 21	2
Pea	7 to 14	3
Pepper	10 to 14	2
Radish	5 to 7	5
Spinach	7 to 14	5
Squash	7 to 14	4
Swiss chard	7 to 14	4
Tomato	7 to 14	4
Turnip	7 to 14	4

Making newspaper pots

Newspaper pots are an inexpensive, biodegradable option for growing and transplanting seedlings. They're also easy to make.

YOU'LL NEED:

- Black and white newspaper (no colored ink)
- Empty 6 oz. tomato paste can, with one end removed
- Scissors

STEPS:

1. Cut the newspaper into strips, 5 to 6 inches long by 12 to 13 inches wide. (Each strip should be about 1½ times the length of the can and double the can's diameter.) Cutting all the strips at once will save you time as you go.
2. Place the can lengthwise on a paper strip, with the can's closed end even with the paper's edge and the overhanging paper at the can's open end.

Paper pots provide a good depth for developing roots and decompose easily in the soil.

Roll the paper around the can.

3. Roll the newspaper around the can.
4. Fold and scrunch the overhanging paper into the can, and press the edges firmly on a surface to flatten.
5. Slide the newspaper off the can.
6. Flip the can around, so the closed end is facing down, and insert it back into the rolled paper.
7. Press down firmly to flatten the folded paper at the bottom.
8. Remove the can and your pot is ready to fill with soil.
9. Most newspapers today use environment-friendly, soy-based ink, so the seedlings can be planted in the garden, pots and all. The plants' roots will not be disturbed and the paper pot will decompose within a few weeks. If you are unsure about the ink used in your local newspaper, give the publishers a call.

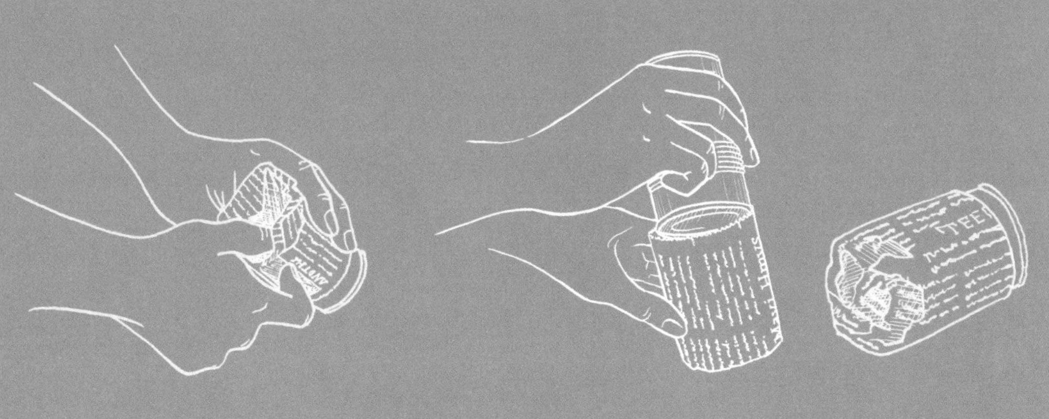

Fold and scrunch the paper inside the can.

Using the other end of the can, flatten the paper to form the bottom of the pot; remove the can and you're ready to plant.

FEBRUARY
Getting Organized and Started

February's main purpose is to test the patience of New York gardeners. During this month, there may be a brief reprieve from the frigid temperatures. Spring bulbs wake up, green shoots start poking through, and you'll be tempted to start working in the garden. But it is always short-lived. Use this time to read about new plants you want to grow, and perhaps stretch your skills and take a couple of local classes. Content yourself with pruning dormant berries and fruit trees, and then turn your attention indoors, where the real action is happening. This is the time to make sure all your seed-starting equipment is assembled and in working order. Tear open your potting mix and inhale the familiar scent of warm, wet soil. Gardeners, get ready to start your seeds!

◀ From planning for spring to starting seedlings, there's lots to do inside while the ground outside is still frozen.

TO DO THIS MONTH

PLAN

- Order or purchase seed
- Organize seed-starting supplies
- Attend garden shows

PREPARE AND MAINTAIN

Everyone

- Dormant prune berry bushes and apple trees
- Check fruit trees for damage and for signs of overwintering insects
- Check on vegetables in storage
- Prepare cold frame

Zones 5, 6, and 7

- Check compost; if it's not frozen, turn the pile and spread finished compost on beds
- Amend soil in garden, as necessary
- Allocate space in the garden for perennial vegetables
- Remove mulch from strawberries

SOW AND PLANT

Zones 5 and 6

- **Sow indoors (late in month):** artichokes, celery, leeks, mint, onions, oregano, and thyme

Zone 7

- **Sow indoors (midmonth):** artichokes, broccoli, Brussels sprouts, cabbage, cauliflower, celery, eggplant, fennel, kale, kohlrabi, leeks, lettuce, mint, oregano, peppers, and tomatoes
- **Direct sow (late in month):** peas, radishes, spinach, and Swiss chard
  TIP *Plant only if soil is workable.*

HARVESTING NOW

From storage

- Garlic
- Parsnips
- Potatoes
- Onions
- Squash
- Turnips

From hoop house

- Arugula
- Beets
- Brussels sprouts
- Carrots
- Herbs
- Kale
- Leeks
- Parsnips
- Swiss chard

Zone 3 Zone 4 Zone 5 Zone 6 Zone 7

Cool-Season and Warm-Season Vegetables

Not all vegetables can be planted as soon as the soil can be worked, and not all vegetables enjoy the heat of summer. Many crops thrive within certain temperature ranges only, including cool-season crops and warm-season crops; others bridge the gap.

Cool-season crops, such as kale, broccoli, and chard, prefer the damp weather of spring and fall. Some crops, such as carrots and lettuce, can be injured by frost. They can be planted early, but you'll need to provide some cover or protection if a frost is expected. Warmer temperatures can actually change their flavors and textures, making them bitter and woody. As temperatures climb above 70°F or so, cool-season crops can bolt, which means the plant stops growing and puts its energy into sending up a seed stalk. During warm months, you can extend the season for cool-season crops by growing them in a shady spot or in the shade cast by corn, tomatoes, or other tall-growing vegetables. Cool-season plants will need more water during hot spells to keep the soil temperature low.

Warm-season crops are frost tender. Even if cold weather does not kill them, it can shock them and stunt their growth. There is nothing to be gained by putting out warm-season crops too early. The plants you hold back until the ground warms up will quickly catch up to earlier seedlings and will probably be stronger and more productive. Wait until after your area's last frost date to plant warm-season crops in the garden. If it's a particularly cool or damp spring, wait even longer. Low temperatures should reliably remain at 50°F or warmer before you plant these crops outside.

▲ Make the most of the growing seasons by using the shade of taller plants to grow cool-loving vegetables such as lettuce, radishes, and spinach.

VEGETABLES BY SEASON

COOL-SEASON

- Asparagus
- Broad beans
- Broccoli
- Brussels sprouts
- Cabbage
- Carrot*
- Kale
- Lettuce*
- Onion
- Parsnip*
- Pea
- Potato*
- Radish
- Rhubarb
- Spinach
- Swiss chard*
- Turnip

WARM-SEASON

- Beans
- Corn
- Cucumber
- Eggplant
- Melon
- Pepper
- Squash
- Tomato

**Cool-season crops injured by frost*

Grow an Indoor Garden for Winter

All kinds of seeds can be started indoors, and although you will be moving most of them out into the garden as soon as the soil is warm enough, you can also start a few plants to continue growing indoors to use throughout the winter. Microgreens and herbs such as chives, basil, parsley, mint, rosemary, and thyme are good candidates for growing on a bright, sunny windowsill. Microgreens are young lettuces and other leafy edibles, such as arugula, beet greens, Asian greens, endive, kale, mustard, and spinach. Use them when they are a couple of inches tall for a tangy topping for salads, soups, and stir-fries. You can buy microgreens seed blends, use a mesclun mix, or put together your own blend. Because you will be harvesting microgreens when they are too tiny to recover, they'll yield only one harvest. To keep a steady supply, start new pots of microgreens every week.

Starting seeds indoors

Growing your favorite varieties of vegetables usually means starting some of them from seed, because you can't count on garden centers to offer a tremendously wide selection of seedlings. Starting seeds indoors requires a bit more effort than planting them outside. You must faithfully keep them watered and place them under an artificial light source for most of the day. Some seeds won't sprout without bottom heat from a heating mat. The payoff is well worth the effort, both in getting exactly the flavors you have been longing for and for the indulgence of vegetable gardening in midwinter.

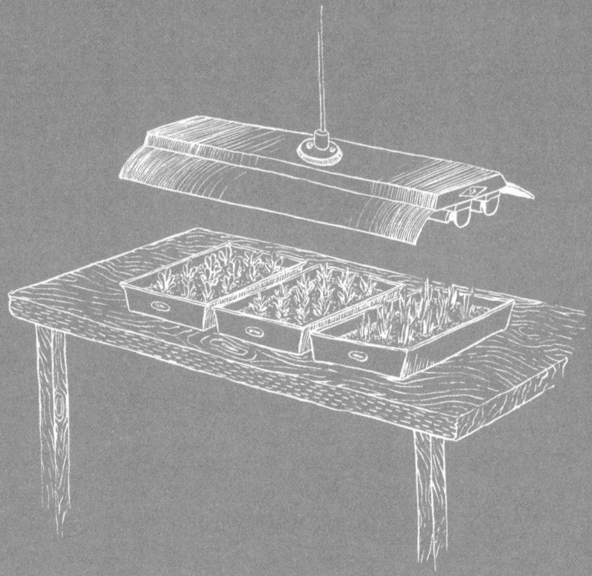

◀ For strong seedlings, fluorescent grow lights should be no more than 3 to 5 inches away, which means you'll have to adjust the height of the light or seedlings as they grow.

YOU'LL NEED:

Containers. You'll find many pots and seed-starting kits at garden centers, but you can use various types of containers for starting seeds. Look around your kitchen and recycling bin. Egg cartons, yogurt containers, take-out trays, and paper cups are all popular seed-starting containers. And don't forget that newspapers can be turned into biodegradable pots.

CONTINUED ▶

YOU'LL NEED, continued

Potting mix. Do not use soil from the garden for starting seeds. It can contain fungus spores and other diseases that will attack young seedlings, and it is heavy and compacts when wet, rather than allowing air and water to flow through. A good potting mix, whether purchased or handmade, will contain a soil substitute, preferably ground bark, and most contain peat plus vermiculite or perlite to lighten and aerate the soil. Avoid mixes that contain water-holding materials and fertilizer, which seeds do not need.

Seeds. If you procrastinated making your catalog seed order, most garden centers offer a good assortment to choose from.

Labels or markers. Seedlings can look nothing like their mature counterparts. Label each container with the plant name and the date planted.

Plastic covers or bags. Place these loosely over a pot to hold in heat and moisture.

Light source. Seedlings need bright light. You'll need to provide an artificial light source during the short days of winter.

Heating mat. This is an optional item. Some seeds, such as tomatoes, need warm soil for germination. Rubber heating mats placed under the seedling containers or trays keep the soil at a constant temperature.

STEPS:

1. It is easier to dampen your potting mix before you put it into the containers, rather than adding water after they are filled. Either place the mix in a large container or on a tarp, and slowly add water, mixing it in until it reaches the consistency of damp sand. It should not be dripping with water and there should be no dry lumps.
2. Fill the containers with the dampened potting mix, and then tap the containers on the table and gently press the mix down to settle it.
3. Now start planting. Check the seed packet for planting depth and any special instructions. Count out larger seeds to plant, or sprinkle in smaller seeds. Either way, add about two or three seeds per container. You can always thin the plants later if you need to.
4. Cover the seeds with more of the damp potting mix. If the mix has dried out, use a very diffused watering can rose or a fine spray to sprinkle the top of the soil while not disturbing the seeds.

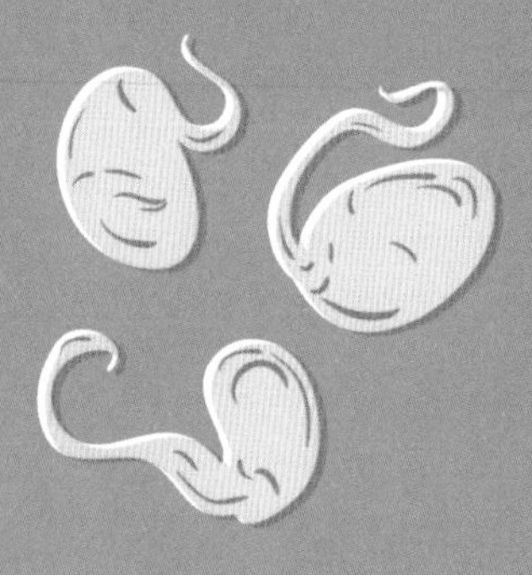

5. Place the containers in a warm spot, 65 to 75°F. They do not need light at this point.
6. Loosely cover the containers with plastic wrap or bags. If you're using seed-starting trays, top them with their plastic covers.
7. Check the soil daily to ensure that it stays moist. If you see mold forming on top of the soil, remove the plastic covering and allow some air to circulate.
8. As soon as you see a green sprout, remove the plastic and provide a source of light. Seedlings will need 12 to 18 hours of light each day. Because we don't get that much daylight during winter and typical home lighting doesn't offer the full light spectrum that plants need, you'll need to supply some type of supplemental lighting. You can purchase special plant or grow lights, but a less expensive option is an ordinary shop light with one warm and one cool fluorescent bulb, or full-spectrum fluorescent bulbs. Fluorescent lights need to be close to the seedlings, ideally within 3 to 5 inches. You can either hang the lights from adjustable chains or cords or prop the seedlings on boxes or boards, removing or lowering them as the plants grow. If you use grow lights that give off heat, they should be a little farther away from the plants—about 4 to 6 inches. Then do yourself a huge favor and use an automatic timer to turn the lights on and off.
9. When you see the first true leaves (they resemble the leaves of the mature plant), you can start feeding the plants. (The first leaves to appear are not actually leaves at all. Those elongated, oval leaves are actually part of the seed and help feed the seedling until true leaves appear.)
10. Depending on the size of the containers and the length of time before you'll be transplanting seedlings outside, you can either leave the seedlings in their original containers or pot them up into something larger. If too many seeds have sprouted in a single container, thin seedlings to the healthiest one and allow it to grow without being crowded. You can repot the extra seedlings, but cutting them off at the soil line is the safest way to avoid disturbing the roots of the seedling you are keeping.
11. Keep the seedlings watered and under the lights. Feed them with your favorite liquid organic fertilizer at about a quarter the rate and dilution recommended for mature plants.

MARCH

Moving Out

They say April is the cruelest month, but March can't be far behind. Spring weather in New York is unpredictable, so you might need to cover and uncover your plants multiple times. Be prepared. The soil should be starting to thaw for most locations, and the first scent of spring—mud—will fill the air. Don't be too eager to start turning soil. You need to let things dry out and warm up. But on the plus side, the days are getting noticeably longer and garden centers are filling up with cool-season plants and supplies. It's also garden show season, a time to indulge in a little garden fantasy and some totally acceptable impulse buying. At the very least, you can check out this year's hottest new gardening gadgets.

◂ Embrace the mud of March. Or at least use it to determine when your garden is ready for planting.

TO DO THIS MONTH

PLAN

- Attend garden shows and classes
- Finalize your garden plans

PREPARE AND MAINTAIN

Everyone

- Check compost; if it's not frozen, turn the pile and spread finished compost on beds
- Amend garden soil, as necessary
- Put up fencing
- Allocate space for perennial vegetables

Zones 3 and 4

- Continue dormant pruning berry bushes and apple trees

Zones 5, 6, and 7

- Test soil in vegetable garden
- Remove mulch from strawberry beds
- Move artichoke seedlings to protected spot outdoors to vernalize
- Trim tops of seedling onions to about 3 inches

SOW AND PLANT

Everyone

- Pot up indoor seedlings as true leaves appear

Zones 3 and 4

- **Sow indoors (late in month):** artichokes, celery, leeks, mint, onions, oregano, and thyme

Zone 5

- **Sow indoors (early in month):** broccoli, Brussels sprouts, cabbage, cauliflower, and kohlrabi
- **Sow indoors (late in month):** basil, celery, chervil, chicory, eggplant, fennel, kale, lettuce, mint, okra, oregano, parsley, peppers, summer savory, and Swiss chard

Zone 6

- **Sow indoors (early in month):** basil, broccoli, Brussels sprouts, cabbage, cauliflower, celery, kale, kohlrabi, mint, okra, oregano, and Swiss chard
- **Sow indoors (late in month):** chervil, chicory, eggplant, fennel, lettuce, parsley, peppers, summer savory, and tomatoes
- **Direct sow:** beets, carrots, green onions, lettuce, peas, radishes, spinach, and turnips
  TIP *Plant only if soil is workable.*
- **Plant outdoors:** asparagus, horseradish, potatoes, and rhubarb

SOW AND PLANT, cont.

Zone 7

- Chit (pre-sprout) potatoes and prepare sweet potato slips
- **Sow indoors (early in month):** chicory, endive, lettuce, and okra
- **Sow indoors (late in month):** basil, cucumbers, melons, parsley, pumpkins, and squash
- **Direct sow (early in month):** arugula, Asian greens, carrots, cilantro, fava beans, onion sets, peas, radishes, spinach, and turnips
- **Direct sow (late in month):** beets, endive, green onions, kale, lettuce, parsnips, and Swiss chard
- **Transplant outdoors:** broccoli, Brussels sprouts, cabbage, cauliflower, kale, leeks, lettuce, onions, oregano, and Swiss chard
- **Plant outdoors:** asparagus, berry bushes, fruit trees, horseradish, potatoes, and rhubarb

HARVESTING NOW

From storage

- Garlic
- Parsnips
- Potatoes
- Onions
- Squash
- Turnips

From hoop house

- Arugula
- Beets
- Brussels sprouts
- Carrots
- Herbs
- Kale
- Leeks
- Parsnips
- Swiss chard

Zone 3 Zone 4 Zone 5 Zone 6 Zone 7

Preparing Beds and Sowing Seed

Preparing to sow seeds outdoors gives you the chance to get to know your soil intimately, because you'll need to get the seed bed ready to support the plants before you sow. Unless your soil is very compacted, you should not need to till an existing planting bed, but you will want to add some amendments to beef up the organic content and nutrients. Start by applying 2 to 4 inches of compost, along with a scattering of organic fertilizer, to the planting areas. Then turn the compost and fertilizer into the top 8 to 10 inches of soil. The compost will improve the soil's moisture retention and provide a welcoming environment for beneficial microbes. Organic fertilizer is slow to release its nutrients, which will still be in the soil when your plants start growing. Finally, use a rake to break up any large soil clumps and make the soil fairly level.

SOWING SEED

Check the back of each seed packet for instructions on planting depth and spacing. Generally, a good planting depth for seeds is twice their diameter. (The diameter of a pea seed, for example, is about ½ inch, so it should be planted about 1 inch deep.) Many vegetable seeds are much smaller than peas and are more difficult to handle. Small seeds should barely be covered with soil. Rather than digging holes or furrows, lightly indent the planting area and do your best to sprinkle the seed evenly. Then top with a light dusting of loose soil and gently tamp the soil. Ideal spacing is next to impossible with tiny lettuce, radish, and carrot seeds. Don't worry about spacing at this point, because you can thin out the plants later and eat the little seedlings.

If you've decided to plant in rows, dig a furrow the desired length and evenly space the seeds in it, and then cover them with soil. For wide rows, plant closely spaced multiple rows or use a hoe to pull back the top few inches of soil, space out the seeds, and sprinkle the soil back on top.

Immediately after planting the seeds, you should do two things: add labels and water. Label your rows so you will know what you planted and, more important, where you planted. You don't want to dig up a previously seeded area or accidently plant carrots on top of peas. Water is the magic elixir of the garden. It is necessary for the seeds to germinate and it keeps them growing. Use a gentle spray of water on your new beds so you don't dislodge or wash away the seeds. Keep the beds damp, but not soaking wet, until the seeds sprout.

After the plants reach 1 or 2 inches tall, you can thin them to their ideal spacing. Be merciless now, because crowded plants will

A shovel test can help determine whether soil is ready to be worked.

compete with each other and won't grow or produce as well as they should. The young seedlings don't have an extensive root system, but pulling them up can dislodge the plants you want to keep. To avoid this, use scissors to snip the excess plants at the soil line. Then use these tiny lettuce, beet, radish, and carrot thinnings in a salad.

Quick Frost Protection

Spring weather is fickle here. The sun may shine for a week or two, and then a surprise overnight frost (or worse) descends upon our unsuspecting spring gardens. Baby plants can't be left out unprotected. A few precautions must be taken, such as installing simple row covers and quick-and-easy hoop tunnels, to cover tender plants and extend the growing season.

ROW COVERS

By far the fastest and easiest way to protect your garden from an impending frost is to toss a row cover over your plants. You can create row covers, or floating row covers, out of just about anything, from a layer of newspapers to an old sheet. Or you can purchase garden fabric created for just this purpose; it offers a few advantages over other types of covers.

Garden fabric is sold in different weights, for different uses. The lightest weight, often called summer weight, is used to protect plants from insects. It can remain in place during the growing season and allows water and light to get through. For cold protection, the best option is a general-purpose garden

HOW TO DETERMINE IF THE SOIL IS WORKABLE

Certain phrases in gardening, such as "well-drained, moist soil" and "as soon as the soil is workable," can leave you scratching your head. Many cool-season fruits and vegetables are labeled with this ambiguous planting direction. "Workable" does not mean "as soon as it has thawed enough to dig into." After a harsh New York winter, soil needs a chance to dry out and warm up. Because winter weather here can vary greatly, a calendar date cannot be set for this phenomenon, as is done for approximating the last spring frost. Some gardeners have come up with an easy, portable test for determining whether the soil is warm and dry enough to plant. They say if you can sit on the soil comfortably, it is ready to be planted. If, however, sitting on damp soil is not how you'd like to inaugurate your gardening year, you can do two tests while standing up.

Shovel test. If you can easily slice your shovel into the ground and it comes out fairly clean, the soil is ready to be worked. On the other hand, if it comes up clumped with mud, you'll need to wait.

Crumble test. Grab a handful of soil and squeeze it into a ball. If it crumbles with a gentle touch of your finger, it is ready to plant. If breaking up the ball of soil requires more force, your soil is too wet for planting. If the soil will not form a ball at all, you can plant, but your garden badly needs water!

fabric, which is a thicker weave that holds in heat, or a heavy-duty garden fabric, for maximum cold protection. The fabric traps the sun's heat, warming the soil and keeping the temperature around the plants about 2 to 5 degrees warmer than the outside air. These covers can prevent frost damage in temperatures down to about 28°F. You can double the fabric for even better protection. They are light enough to avoid crushing or snapping off tender plants, or you can lay them over wire hoops or some other structure to keep them off the plants. If you leave the hoops in place, you can add the row covers for protection at the end of the season, too.

HOOP HOUSES

A hoop house is a portable greenhouse you can make by arching PVC pipe into a frame and covering it in plastic or ventilated garden film. Hoop houses can provide more reliable cold protection than row covers, plus they are easy to make. You can build your own hoop house to shelter overwintering vegetables such as late-season kale and carrots.

The temperature inside the hoop house will warm up weeks before the exposed beds in your garden, and this allows you to get a head start on spring planting. As your plants grow under the hoop house, keep tabs on the temperature. During warm spells, you may need to open the ends of the hoop house to ventilate and cool the plants inside. After snow covers the hoop house, it may become impractical to continue harvesting, but the plants will be safely tucked away and waiting for you at the next thaw. When you do harvest, wait until the sun has had a chance to warm the plants inside so they are not partially frozen when you bring them indoors. And when you've

Hoop houses can help protect edibles from cold temperatures.

finished harvesting, be very sure to secure both ends of the hoop house.

You can take down the hoop house at the end of winter or remove the covering and leave the hoops in place and handy for the next fall. Gardeners in zones 3 to 5 can leave the cover on, with ventilation, and grow enviable hot-season prizes such as eggplants and peppers a bit earlier than usual. Or cover the hoops with shade cloth to grow cool-season crops throughout the summer.

Building a hoop house

By arching PVC pipe across a prepared bed and covering it with heavy plastic or garden film, you can create a 6-by-3-foot hoop house to cover your plants and protect them throughout the winter months.

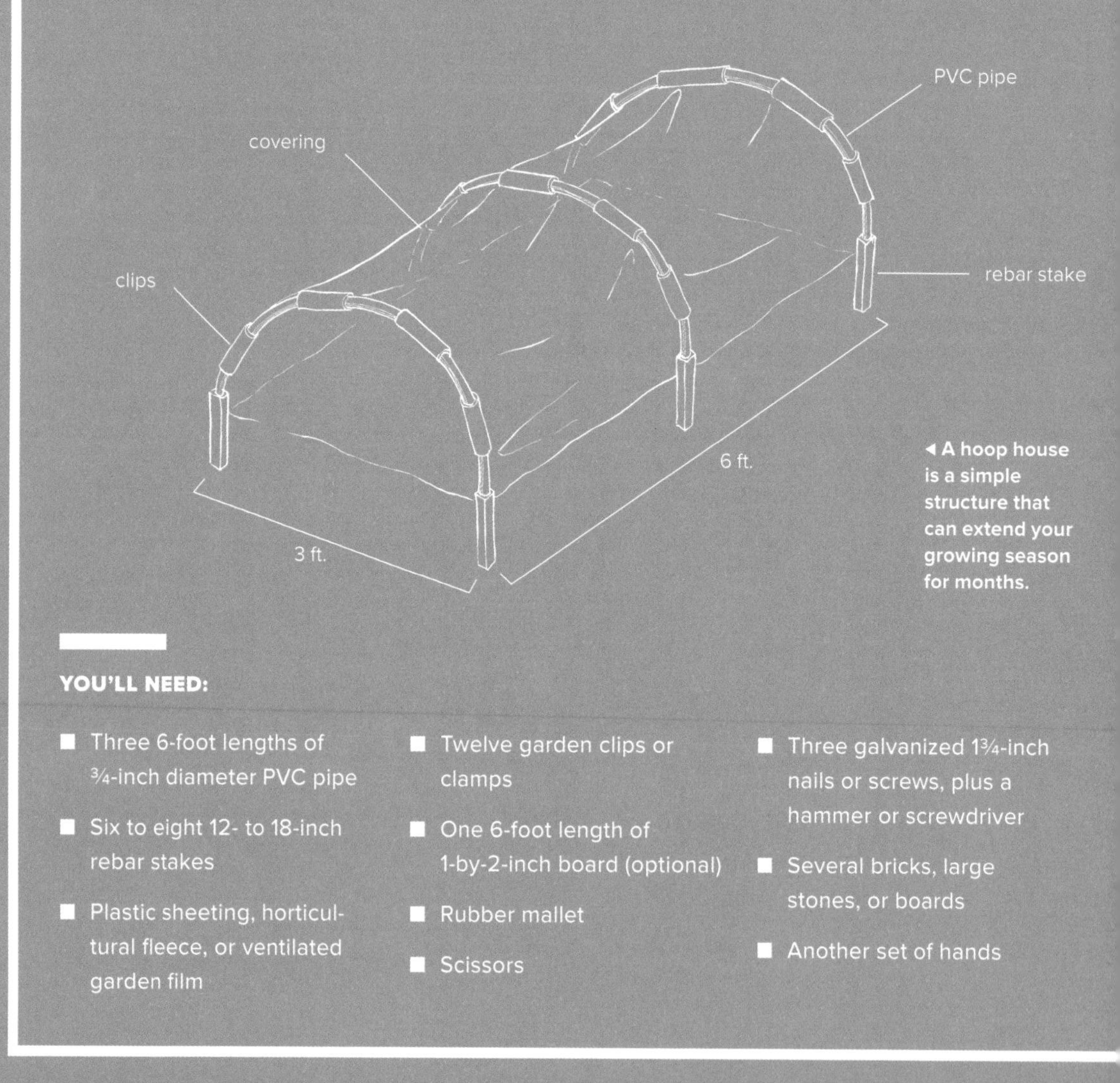

◀ **A hoop house is a simple structure that can extend your growing season for months.**

YOU'LL NEED:

- Three 6-foot lengths of ¾-inch diameter PVC pipe
- Six to eight 12- to 18-inch rebar stakes
- Plastic sheeting, horticultural fleece, or ventilated garden film
- Twelve garden clips or clamps
- One 6-foot length of 1-by-2-inch board (optional)
- Rubber mallet
- Scissors
- Three galvanized 1¾-inch nails or screws, plus a hammer or screwdriver
- Several bricks, large stones, or boards
- Another set of hands

BUT FIRST:

Before you start a hoop house project, consider a few factors:

- Prepare the planting bed before erecting the hoop house.
- Even easier, build the hoop house to cover an existing raised bed.
- Make sure you can reach into the center of the bed to harvest crops with the hoop house in place.
- To gain the most benefit from your hoop house, situate it where it will receive the most sunshine.
- Do not put your hoop house at the bottom of a slope, where cold air can collect and spring runoff could flood it.

STEPS:

1. Pound a rebar stake halfway into the ground at each corner of the bed and on either side of the center of the bed.
2. Slip one end of a PVC pipe over the exposed portion of rebar, and very, very carefully bend the PVC to the rebar stake on the opposite side of the bed. Repeat with the remaining pipes.
3. For additional support, you can lay the 6-foot length of 1-by-2-inch board over the center top of the curved PVC and nail or screw it to the pipes.
4. Drape the plastic, greenhouse poly, or garden film evenly over the frame. Secure it to the frame with the garden clips or clamps to prevent it taking flight in the wind.
5. Cut away any excess material along the sides, leaving enough overhang to weigh down and secure with bricks, stones, or a board. You could also shovel soil over the ends, but soil is not as convenient to remove.
6. Close the end opening of the hoop house by twisting the covering tightly and securing it in place by tying it to additional rebar stakes or by weighing it down with several heavy bricks.
7. Double-check that everything seems secure. Winter winds can overturn light-weight hoop houses, and it is no fun trying to get them back in place with frozen hands!

APRIL
Friend or Foe?

Spring is here, and the warmer temperatures lure us outside. Whether you have already started planting or are checking for signs of life from the spinach and mache seeds you put out last fall, April moves quickly. The warmer weather also means that you can safely begin transplanting seedlings outdoors and into the garden. By the end of this month, everyone will be in full gardening mode. Planting may be our main focus, but gardeners are not the only busy creatures outside. The animals are waking up, and they are hungry. The insects are hatching and flying in. Tiny, tender seedlings could easily fall prey to a miscellany of pests. There will be no rest for the gardener in April.

◀ Plant a garden hospitable to ladybugs and other beneficial insects and they'll help keep your harvest pest free.

TO DO THIS MONTH

PLAN

- Make notes in your garden journal
- Finalize your garden plan

PREPARE AND MAINTAIN

Everyone

- Fertilize garlic and shallots
- Continue to pot up seedlings as true leaves appear
- Keep those seedlings watered
- Set up trellises and other supports

Zones 3 and 4

- Move artichoke seedlings to protected spot outdoors to vernalize

Zones 5, 6, and 7

- Prune apricot, cherry, and peach trees
- Keep an eye out for cutworms, flea beetles, and leaf miners
- Thin beets, carrots, and lettuce, and eat the thinnings
- Chit (pre-sprout) potatoes and prepare sweet potato slips
- Begin mulching the garden
- Divide rhubarb

SOW AND PLANT

Zone 3

- **Sow indoors (early in month):** basil, broccoli, Brussels sprouts, cabbage, cauliflower, celery, kohlrabi, and peppers
- **Sow indoors (late in month):** chervil, chicory, eggplant, endive, fennel, kale, lettuce, mint, okra, oregano, parsley, summer savory, Swiss chard, and tomatoes
- **Direct sow (late in month):** arugula, Asian greens, beets, broccoli raab, carrots, cilantro, endive, fava beans, green onions, lettuce, onion sets, parsnips, peas, radishes, spinach, and turnips
  TIP *Plant only if soil is workable.*
- **Plant outdoors:** asparagus, berry bushes, fruit trees, horseradish, potatoes, and rhubarb
- **Transplant outdoors (late in month):** celery, leeks, mint, onions, oregano, and thyme

Zones 4 and 5

- **Sow indoors (early in month):** chicory, endive, and lettuce
- **Sow indoors (midmonth):** cucumbers, melons, okra, pumpkins, squash, and tomatoes
- **Direct sow (early in month):** arugula, Asian greens, cilantro, endive, fava beans, peas, and spinach
- **Direct sow (late in month):** beans, beets, broccoli, carrots, green onions, lettuce, onion sets, parsnips, radishes, Swiss chard, and turnips

SOW AND PLANT, cont.

- **Transplant outdoors (early in month):** artichoke, broccoli, Brussels sprouts, cabbage, cauliflower, celery, chicory, collards, fennel, kale, kohlrabi, leeks, lettuce, mint, onions, oregano, parsley, Swiss chard, and thyme
- **Plant outdoors:** asparagus, berry bushes, fruit trees, horseradish, potatoes, and rhubarb

Zone 6

- Harden off remaining seedlings
- **Sow indoors (early in month):** chicory, cucumbers, endive, lettuce, pumpkins, and squash
- **Direct sow (early in month):** arugula, Asian greens, carrots, cilantro, endive, fava beans, onion sets, parsnips, peas, radishes, and spinach
- **Direct sow (late in month):** beans, beets, broccoli raab, cabbage, cauliflower, green onions, lettuce, and turnips
- **Transplant outdoors:** artichoke, broccoli, Brussels sprouts, cabbage, cauliflower, celery, fennel, kale, kohlrabi, leeks, lettuce, mint, onions, oregano, parsley, Swiss chard, thyme, and tomatoes
- **Plant outdoors:** asparagus, berry bushes, fruit trees, horseradish, potatoes, and rhubarb

SOW AND PLANT, cont.

Zone 7

- Harden off remaining seedlings
- **Direct sow (early in month):** broccoli, broccoli raab, cabbage, cauliflower, cilantro, fava beans, lettuce, and rutabaga
- **Direct sow (late in month):** beans, corn, cucumbers, melons, okra, pumpkins, rutabaga, and squash
- **Continue succession planting:** beets, broccoli, carrots, green onions, kale, and lettuce
- **Transplant outdoors:** artichoke, basil, celery, chicory, cucumbers, eggplant, endive, fennel, kohlrabi, melon, mint, okra, oregano, parsley, peppers, pumpkins, squash, thyme, and tomatoes
- **Plant outdoors:** asparagus, berry bushes, fruit trees, horseradish, potatoes, and rhubarb

HARVESTING NOW

- Carrots
- Early greens and herbs
- Lettuce
- Remaining overwintered parsnips
- Wild vegetables: fiddleheads, ramps, and pea tendrils

Zone 3 Zone 4 Zone 5 Zone 6 Zone 7

Transplanting

Whether you started seeds indoors or purchased seedlings, transplanting them outdoors gives you an instant garden—but first you must make sure your garden is ready. Start by preparing your planting beds as you would for direct sowing, preferably a few weeks before you transplant. If you decide to purchase seedlings, look for short, stocky plants with well-developed root systems. Healthy, sturdy plants will suffer less transplant shock than larger plants that are already stressed from growing in a cramped pot. Avoid plants that have set flowers or fruit. A fruiting plant will put its energy into ripening those fruits, not settling in and setting more fruits.

Correctly transplanting containerized plants involves a few easy steps.

1. Dig a hole about the same size and depth as the container. Note: If you're transplanting tomatoes or peppers, these plants can grow roots from their stems, so planting them a little deeper than the pot depth will make them stronger plants. Tomatoes can be planted all the way up to their top set of leaves. Peppers should be planted about ½ inch deeper than they were in their pots.
2. Tap and/or squeeze the container to loosen the plant's roots. Then place your hand over the top of the pot, fingers around the stem, and turn it upside down and ease out the plant. Don't pull on the stem and risk breaking it and killing the plant.
3. If the roots are compacted or pot-bound, gently tease them apart with your hands or a fork.
4. Place the plant in the hole, backfill with soil, and lightly firm it in.
5. Water well and top with a layer of mulch.
6. For plants that will need staking, add stakes now while the roots are small. (If you wait until the plants need staking, you might accidently stab a stake through the spreading roots. Plus it's nice to have stakes in place and ready to use when plants need them.)
7. Label your transplants so you'll know which is which, and then step back and admire your almost instant garden. But do keep the row covers handy, in case the temperature dips.

Deer are beautiful, but they can become pests in the vegetable garden.

Defending Your Garden Against Four-Footed Pests

After your seeds start sprouting and your transplants are nestled safely in the garden, don't assume it's time to rest on your laurels. We humans are not the only animals who enjoy fresh greenery. Even pesky insects can't compete with the destruction caused by four-footed pests in the garden. I am talking about deer, rabbits, groundhogs, skunks, chipmunks, squirrels, and cats and dogs—to name a few. Animals can clean off a berry bush or turn a vegetable garden into stubs in a matter of hours. It is virtually impossible to grow edible plants in the Northeast without a fence, and even that might not be enough to keep all the animals out. Before filling your garden with delectables, you can take a few precautions against marauding wildlife. Keep in mind, however, that animals tend to wise up after a while and will find a way around almost any deterrent. Your best bet is to combine or alternate several methods. And remember that no deterrent is going to stop a starving animal.

PHYSICAL BARRIERS

This is deer country, or so the little dears seem to think. A good fence is well worth the installation effort and expense. You'll find all kinds of resources for building a deer fence, and a great place to start is at your local cooperative extension office. As you're considering fencing, keep three things in mind: how high, how low, and mesh size.

Smaller gardens need a fence at least 8 feet tall to deter deer from jumping over it.

HARDENING OFF SEEDLINGS

Although it's tempting to move your plants out into the garden on the first warm day, tender seedlings need time to acclimate to the change in temperature, the wind, and sun exposure. This is called hardening off. You can choose from a couple of methods for hardening off your plants; both take a week or two.

Expose your plants gradually to longer and longer periods of outdoor, open exposure. A week or two before you plan to transplant, move seedlings to a shaded outdoor spot, away from strong winds. Leave them outdoors for 3 or 4 hours and then move them back indoors. Increase the time outdoors by 1 to 2 hours each day, adding a few hours in direct sunlight after the third or fourth day. After 7 to 10 days, the seedlings should be ready to stay out all day and night and can be safely transplanted into the garden.

Use cold frames to make hardening off even easier. Starting a week or two before your transplanting date, move your seedlings into the cold frame. Open the cover for 3 or 4 hours at first, and gradually increase the time open by 1 to 2 hours a day until the cover can remain open or off all day and night.

▲ Cold frames can help ease the hardening off process.

A FEW HANDY TIPS

- Whichever method you use, you'll need to work with weather changes and be prepared to cover and protect the plants, or move them back indoors, if nighttime temperatures threaten to fall below 40°F.
- It's easier to move pots if you put them on wheels. A wheelbarrow or wagon can be rolled in and out of a garage or shed with ease.
- Don't let the soil get too dry. Wind can dry out the soil in tiny pots in a matter of hours.
- Seedlings are easy prey for hungry animals and insects, especially snails and slugs. Protect the plants with appropriate barriers.
- When you are ready to transplant, water the seedlings in the pots, and then transplant them in late afternoon, in the evening, or on an overcast day to allow the plants to settle in before being exposed to heat and sun.

It doesn't have to be an expensive, sturdy fence—even netting at that height is enough to make a deer think twice before leaping. Smaller pests, such as groundhogs and rabbits, will chew through flimsy fences or burrow under tall ones. To foil these pests, you will need to bury 10 to 12 inches of sturdy fencing material around the entire perimeter. Hardware cloth is an excellent choice, with holes small enough to keep most pests out. Bury the fencing at an outward angle to discourage the diggers even more.

To stop browsing of freestanding berry bushes and fruit trees, netting or caging are good options. You will need to cover the plants entirely and make sure there is no way in, but the covering is needed only while fruit is on the plants. Animals don't usually wait until the fruit is fully ripe to start eating it, so don't delay installing protection.

VISUAL DETERRENTS

Groundhogs, like their squirrel cousins, can also climb. Fencing can go only so far in keeping animals out, so you will need to do more. If you can't block them, scare them. Visual deterrents can include anything from dangling CDs or strips of foil that catch sunlight and randomly flash warnings, to fake snakes and owls with eyes that seem to follow the little thieves, to motion-activated sprinklers that shoot a blast of water when an animal crosses the path of their electronic eyes. These sprinklers are battery operated and easy to install, and some types don't need to be hooked up to a hose.

SOUND DETERRENTS

Animals have much keener senses than we do, and that includes their hearing. You can install your own low-tech sound deterrents by hanging foil strips that rustle in the breeze. Or you can go high-tech by installing deterrents that make the most of the animals' keen hearing by using either audible (sonic) or inaudible (ultrasonic) sound to ward off the intruders. Many are motion-activated to detect when an animal crosses their path. Ultrasonic devices that change and pulse frequencies can be effective for a wide range of animals, including skunks, raccoons, deer, rabbits, rodents, cats, and dogs—all while being virtually inaudible to people.

ODOR REPELLENTS

Two types of scent can be used in the garden to deter animal pests: offensive and predatory. Some animals don't like the smell of particular plants. Interplanting these repellant plants with some of their favorites—abhorrently fragrant lavender alongside tasty Swiss chard, for instance—offers some protection from deer. Predator scents, often in the form of urine, tell animals to steer clear of the area. You can buy predator urine in a powdered form, which does not wash away as quickly as liquid urine. Even the scent of human hair and some perfumed soaps can be enough to tell skittish animals, such as deer and rabbits, that they are not safe here.

TASTE AVERSIONS AND DISTRACTIONS

Coating plants with a taste or flavor they find unappealing, such as hot peppers, castor oil, or garlic, is a good solution for protecting ornamental plants, but it's not as useful on edible plants. The taste will linger, and your family may find it just as unappealing. These coatings are best used on the parts of plants you will not be eating, such as the base of pepper and cucumber plants, but not on lettuce

COMPANION PLANTING

Companion plants help each other in the garden. Good plant companions can deter pests, attract beneficial insects, and even improve growth and yields. Companion planting is a long-established gardening technique that takes advantage of the best pairings to the benefit of the plants. It can get complicated trying to match up plants, but if you experiment a little, you'll find companion planting to be a great tool in your battle against bugs.

An easy place to start is by incorporating herbs in your vegetable garden to deter some of the peskier pests. You can try a few things that will convince you of companion planting's appeal.

PESTS AND DETERRENTS

To deter	Plant
Aphids	Chives and coriander
Bean beetle	Marigolds, nasturtiums, and rosemary
Cabbage moths	Mint, oregano, and sage
Carrot flies	Rosemary and sage
Flea beetles	Catmint and mint
Squash bugs and beetles	Nasturtiums and tansy

and kale leaves. Note that before you apply anything directly to your plants, you'll need to be certain it is approved for use on edible plants.

I've had good luck diverting rabbits and groundhogs from my coveted plants by overseeding my lawn with clover, one of their favorite snacks. I also let the wild violets grow and spread throughout the yard. Because the animals are kept well fed elsewhere in the yard, they don't focus on my garden.

Slug and Snail Control

Slugs. Even the name sounds disgusting. Slugs and snails love the tender plants and damp soil of vegetable gardens. They can devour entire seedlings and leave mature plants riddled with holes or chewed down to skeletons. They must be stopped, and you can use several approaches for keeping them under control.

Removal by hand. This is straightforward enough: When you see a snail or slug, pick it up and remove it. If you are too squeamish to squish them, you can toss them into a bucket of salty water or move them to an exposed sidewalk for the birds to enjoy.

Predators. Slugs have natural predators that you should welcome into your yard and garden. Birds and frogs are good garden partners, and chickens and ducks can clear a yard of snails and slugs in no time.

Barriers. It would be nice if we could keep slugs from getting into our gardens in the first place, but a fence definitely won't do the

Remove slugs and snails by hand if you see them.

trick. And even though several herbs and plants repel slugs—including chives, fennel, foxgloves, garlic, mint, sage, and sunflowers—plants aren't enough to stop snails and slugs; you'll also need to use slug barriers that exploit their soft underbellies. Any substance with sharp edges is lethal to slugs, cutting their undersides and causing dehydration. Eggshells and used coffee grounds are readily available. Diatomaceous earth (DE), the fossilized remains of diatoms (an algae), is considered an excellent organic slug control; it works best if it remains dry. (Just be sure not to inhale it while applying it, and always purchase food-grade or DE labeled for pest control—not the product sold in pool supply stores.)

Copper strips can be used as barriers. When a slug slithers over copper, its moist body reacts with the metal, causing a tiny electrical shock; this doesn't kill the slug, but it certainly makes him think twice about crossing it. The strips need to be about 2 or 3 inches wide to prevent slugs from lifting themselves over it.

Traps. Slugs love cozy, dark corners and will seek out any cover, such as boards, upside down pots, or old cabbage leaves. Leave some of these strategically placed in your garden and you will find colonies of slugs to dispose of in the morning. You can use baits to lure slugs into the traps. Dry pet food under an overturned pot works well, and beer is a classic slug bait. Sink a small container, such as a yogurt cup, into the soil so that the top edge is at ground level. Half fill it with beer and the slugs will dive in and be unable to get out. Increasingly popular are nontoxic organic baits made with iron phosphate; sprinkle them around plants to attract snails and slugs. When ingested, the baits cause slugs to stop feeding and retreat underground, where they die. These baits do not harm your plants, the soil, other insects, or animals and humans.

Beneficial Insects

Not all insects are bad news. Gardeners tend to think of insects as the enemy, but many insects are just hanging out, and some are actually our allies in the garden. We call the good guys beneficial insects, although I am sure all insects have their purpose somewhere. Some beneficials will gobble down more aphids, thrips, slug eggs, and assorted caterpillars than you even realized were in your garden. Others are necessary to help pollinate our plants for an abundant crop.

It's so important not to reach for the bug spray every time you see a potential problem. Most commercial insecticides kill a broad spectrum of insects, good and bad. We want to achieve a balanced ecosystem in our gardens, and that means encouraging beneficial insects to help control the levels of pest insects and keep our soil rich and alive. Before insect pests become a problem, you can encourage more beneficial insects into your garden by using a few techniques and tactics.

PLANT TO ATTRACT

Include flowers and herbs that attract adult beneficials. Hover fly and lacewing larvae will feast on aphids and caterpillars, respectively, but the adults like to dine on pollen and nectar. Flowers in the garden lure in the adults, and hopefully they will lay their eggs in the vegetable patch. Small hover flies and lacewings flock to umbels—plants with umbrellalike flowers, including dill, fennel, and parsley. The tiny flower clusters are perfectly proportioned for these insects.

Larger beneficials, such as ravenous ladybugs and predatory wasps who lay their eggs inside the tomato hornworm, prefer flowers in the composite family, such as coreopsis, cosmos, sunflowers, and zinnias. Ladybugs are particularly fond of goldenrod. You can keep ladybugs in the area by occasionally spraying plants with a sugar-water mix. Dissolve 5 ounces of sugar in a quart of warm water. Allow this to cool and then spray wherever aphids are a problem. The ladies should find it quickly.

The humble ground beetle feeds after dark, snacking on caterpillars, cutworms, maggots, and their eggs. To make them feel at home, provide low-growing cover for daytime hideouts with spreading herbs, such as oregano, thyme, and savory. Even a nice, damp layer of mulch will do.

PROVIDE WATER OR PUDDLES

Insects get thirsty, too. They aren't particular about where their water comes from, but if you use drip irrigation or soaker hoses, they may not have access to the water they need. A shallow dish of water among the herbs and flowers may mean the difference between resident beneficials and fly-bys.

GIVE IT TIME

Exercise some patience. It takes longer for a garden to build up a population of beneficials than to become overrun with pests. But unless the good guys find something to eat, they won't visit your garden. If you can discipline yourself to leave that banquet waiting for them, they will come and invite their friends. Once beneficials find your garden, it won't take long before the pest problem disappears.

Homemade pesticides

There will be times when you'll need to deal directly with pest insects. If loopers are turning your cabbages into filigree, don't feel guilty about stopping them. Always reach for the least toxic solution first, however, such as removing by hand, adding a row cover, or applying *Bacillus thuringiensis* (Bt). Reaching for the pesticide can is always a last resort in the vegetable garden, even when the pesticide is an organic control. If insects and diseases gain a foothold and you need to regain control, you can mix up your own solutions in your kitchen and feel good about using these safe mixtures on edible plants. That doesn't mean you should use them with abandon, however—even homemade can be overdone.

When you do need to spray, be sure to wet the undersides of leaves, where insects hide and lay their eggs. And always be careful when spraying any type of pesticide; keep it out of your eyes and off your skin.

INSECTICIDAL SOAP

One of the safest pesticides, soaps are effective against aphids, chiggers, earwigs, mealybugs, mites, thrips, and whiteflies. The fatty acids in soap penetrate the outside of soft-bodied insects and kill them by collapsing their cell membranes. It must be sprayed directly on the insect and is not effective after it has dried on the plant. Soap can be used right up to harvest time, but keep in mind that it can burn the leaves of some plants. To be safe, you can hose it off after a few hours, and do not use it in full sunlight.

Ingredients

- 2 Tbsp. baby shampoo
- 1 gallon water

To use: Mix well and spray on insect pests.

CONTINUED ▸

HORTICULTURAL OIL

Oils are effective against a wide variety of insects, including those affected by insecticidal soap, plus caterpillars and leafhoppers. They can also help control powdery mildew. Like soap, oils are nontoxic and safe to use on edible plants. Spray them on the insects, and they coat and suffocate them and penetrate their eggs. Because oil does not mix easily with water for application, you can add a little soap to the solution to help the oil emulsify and spread evenly when sprayed. Note that oils can quickly burn plant leaves and should be hosed off after a few hours.

Ingredients

- 1 cup vegetable oil (cottonseed oil and soybean oil are recommended)
- 1 Tbsp. liquid soap (dishwashing liquid or baby shampoo)
- 1 gallon water

To use: Mix the oil and soap together well and then add to the water. Spray directly on insect pests.

GARLIC OIL

Chemical compounds in garlic repel or kill many insects, including aphids, whiteflies, and many beetles. Use it sparingly, however, because it repels insects indiscriminately, even those you need for pollination, and you don't want to chase every insect out of your garden. Because the mixture contains oil, don't use it in full sunlight or it will burn the plants. The garlic odor can linger for a week or more and will continue to repel insects.

Ingredients

- 3 or 4 garlic cloves, minced
- Orange peel (optional)
- 2 tsp. mineral oil
- 2 cups water
- 1 tsp. dishwashing liquid

To use: Mix the minced garlic and mineral oil together and allow to sit for 12 to 36 hours. Strain or squeeze the garlic from the oil, and then mix the oil with the soap and combine with the water. For an extra whammy, you can add orange peel to the minced garlic and mineral oil. Store the mixture in the refrigerator in a tightly sealed container for up to a week.

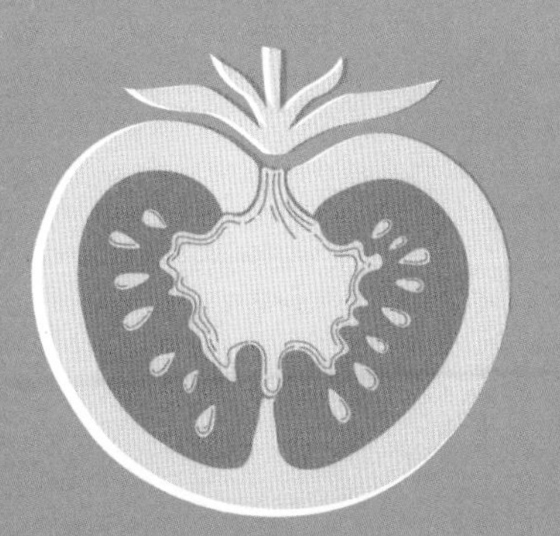

TOMATO/POTATO LEAF SPRAY

All members of the nightshade family contain toxic alkaloids in their leaves. You can chop up the leaves, add them to water, let it steep, and then spray it on plants to repel or kill aphids and corn earworm. Although these alkaloids repel some pests, they actually attract beneficial insects.

Use this spray with caution: Avoid getting it in your eyes and do not use it near people who are allergic to tomatoes or other members of the nightshade family. To avoid spreading the mosaic virus (which can reside in the leaves), do not use the spray on other members of the nightshade family, including peppers and eggplants.

Ingredients

- 1 to 2 cups tomato or potato leaves, finely chopped
- 4 cups water

To use: Soak the chopped leaves in half of the water overnight. Strain, and add the liquid to the remaining water. Spray directly on plants.

MILK SPRAY FOR POWDERY MILDEW

Powdery mildew is a fungal disease that coats leaves in powdery, white splotches. It won't kill an infected plant, but it can impair photosynthesis and diminish the flavor of the fruit. If you catch it early enough, you can control it by applying a dilution of ordinary milk. Unlike oil-based cures, this mixture must be applied while the sun is shining. Some gardeners complain about the smell of spoiled milk; you can dilute the recipe a bit if that's a problem.

Ingredients

- 3 to 4 parts milk (low fat is fine)
- 6 to 7 parts water

To use: Mix the ingredients together and thoroughly saturate the plant leaves on both sides. The mixture will kill any existing fungal spores, but you will need to reapply it every 10 to 12 days to prevent a reoccurrence.

MAY

Pushing the Limits

May could be the most pleasant month to be in the garden. The sun reestablishes its dominance. The days slowly lengthen. The wind gives way to a breeze, picking up the fresh scent of unfurling leaves. The plants are showing the effect of spring rains and warming temperatures. Fruit trees are so heavy with blossoms that we celebrate them with festivals. Of course, the insects are back, but that's not all bad. By now, we're enjoying eating from the garden and planting is in full swing. Successions of cool-season vegetables, such as arugula, beets, radishes, lettuce, peas, and spinach are turning salad into a banquet. But their days are numbered, and it's time to move the sun worshipers outdoors and get the garden set up for the sprint into summer.

◀ There's a lot to do as summer approaches. Enlist some help and get your warm-weather seeds in the ground.

TO DO THIS MONTH

PLAN

- Make notes on the weather, which vegetables did well, and what to plant more of next spring
- Double-check that you have all the seeds you need

PREPARE AND MAINTAIN

Everyone

- Shop for transplants
- Weed and mulch
- Start a new compost pile
- Scout for signs of insects and eggs
- Cover seedlings that need protection from early season pests
- Keep row covers handy for late spring frost warnings

PREPARE AND MAINTAIN, cont.

Zones 3 and 4

- Test soil in vegetable garden
- Prune apricot, cherry, and peach trees
- Keep an eye out for cutworms, flea beetles, and leaf miners
- Thin beets, carrots, and lettuce, and eat the thinnings
- Chit (pre-sprout) potatoes and prepare sweet potato slips
- Begin mulching the garden
- Divide rhubarb

Zones 5, 6, and 7

- Thin seedlings of beets, carrots, and salad greens
- Set up trellises for vining crops
- Place cutworm collars on new transplants
- Remove blossoms from new strawberry plants

Zone 3 Zone 4 Zone 5 Zone 6 Zone 7

SOW AND PLANT

Zones 3 and 4

- Repot warm-season vegetables into larger pots
- **Sow indoors (early in month):** chicory, cucumbers, endive, lettuce, melons, okra, pumpkins, squash, and tomatoes
- **Direct sow:** arugula, Asian greens, beans, beets, broccoli raab, carrots, corn, cucumbers, green onions, kale, lettuce, melons, parsnips, peas, pumpkins, radishes, rutabaga, spinach, squash, Swiss chard, and turnips
- **Plant outdoors:** asparagus, horseradish, and potatoes
- **Transplant outdoors:** artichoke, basil, broccoli, Brussels sprouts, cabbage, cauliflower, celery, chicory, fennel, kale, kohlrabi, leeks, mint, okra, onions, oregano, parsley, Swiss chard, thyme, and tomatoes

Zones 5, 6, and 7

- Plant sweet potatoes
- **Direct sow:** beans, broccoli raab, corn, cucumbers, lettuce, melon, okra, pumpkins, rutabaga, and squash
- **Transplant outdoors:** basil, celery, cucumbers, eggplants, endive, kohlrabi, melon, mint, okra, peppers, pumpkins, squash, thyme, and tomatoes
  TIP *Transplant these after the last frost date.*

HARVESTING NOW

- Arugula
- Asparagus
- Broccoli raab
- Fava beans
- Fiddleheads
- Garlic scapes
- Green onions
- Lettuce
- Radishes
- Rhubarb
- Spinach
- Swiss chard

Growing Herbs

No vegetable garden is complete without herbs. Tomato dishes need basil. Mint excites peas. Chives make the baked potato. Herbs don't need a lot of space; you can tuck them in between vegetables or give them a spot of their own. I plant my perennial herbs—sage, lavender, oregano, and thyme—in a corner of my vegetable garden, where they can grow undisturbed. I plant the annuals within the other vegetable beds. Many of the annual herbs attract beneficial insects, so they are doing double duty.

The key is to keep your herbs handy so you'll use them often. When I go to the garden to see what looks good for dinner, the herbs are right there with my vegetables. Most herbs are also ideal for containers and can be grown near the kitchen or even inside on a sunny windowsill.

HERB-GROWING TIPS

Sun is the secret to growing herbs. With the exception of mint, kitchen herbs need 6 to 8 hours of sun a day to thrive and become flavorful. Give them a little water and well-draining soil and they'll take off with the warming weather.

What herbs do not need is lots of fertilizer or an overly rich soil. Although feeding them will give you lush, green growth, growing them on the lean side will concentrate their essential oils and intensify their flavors. You can coax more growth by pinching and snipping (and then using); every time you cut them back, you encourage the plants to branch out and get bushier. One of the biggest mistakes gardeners make with herbs is being afraid to cut them. The best flavor comes from herbs that are actively growing. Left unused, herb plants will flower and go to seed, which can make them tough and bitter, so don't be afraid to indulge and enjoy.

BEST HERBS FOR NEW YORK

These herbs are right at home in a New York garden.

Basil. This fast-growing annual is as aromatic as it is tasty, and it comes in dozens of varieties with flavors from minty, to crisp and citrusy, to warm and spicy.

Bay. This small tree is best grown in a container in New York because it needs to be brought indoors for the winter. Its fresh leaves have a pungent musky, mint flavor, and one leaf will work magic in a recipe. They make excellent houseplants.

Oregano flowers attract pollinators such as bees.

Chives. This perennial herb is one of the first green shoots of spring. Clip entire blades and the plants will regrow. Its delicate onion flavor is best enjoyed uncooked.

Cilantro. Love it or hate it, cilantro needs to be grown quickly in the cool of spring or fall. Its distinct aroma and flavor make it a staple ingredient in many Mexican and Asian cuisines.

Dill. The aroma of this lovely, feathery plant is so strong you will pick up the fragrance from across the yard. Both the delicate leaves and the seeds are used in cooking, and even the flowers are edible.

Fennel. Fennel may look similar to dill, but it's flavor is reminiscent of minty licorice. This biennial goes to seed in its second year. It will readily self-sow, and both the seeds and flowers are edible.

Lavender. In New York gardens, few lavender plants will grow into shrubs, but we can still enjoy the glorious flowers on smaller plants. Both the leaves and flowers are delicious in savory and sweet dishes.

Lemon balm. As a perennial that spreads readily, this mint family plant is an enthusiastic grower. Its wonderfully fresh lemon flavor is used in seasoning and tea.

Mint. The Northeast has a love-hate relationship with mint. Yummy in all its many flavors, including peppermint, chocolate, orange, and pineapple, it can take over a garden quickly. I recommend growing it in a container.

EDIBLE FLOWERS

It turns out you can eat the daisies. You might not want to, though, because there are better tasting flowers. Before you chomp on a blossom, however, make sure the plant has been grown organically, with no pesticides. In that case, here are some delicious choices to try:

- Anise hyssop
- Bean blossoms
- Bee balm
- Borage
- Calendula
- Chrysanthemum
- Daylily
- Dianthus
- Hollyhock
- Lavender
- Marigold
- Nasturtium
- Pansy
- Pea blossoms
- Rose
- Sunflower
- Squash blossoms
- Violet

Oregano. Greek oregano is the variety associated with tomato sauce and Mediterranean cooking, but all varieties grow well here. Bees love the flowers, which is a big plus in the garden.

Parsley. A biennial plant, parsley grows all season without going to seed. Fresh parsley has a grassy, herbal flavor that you won't want to waste as a mere garnish.

Rosemary. This evergreen also needs to be brought indoors for the winter. Its dark green needles and an intense pine-like scent and flavor envelopes the kitchen and is perfect for fall comfort food. A little goes a long way.

Sage. Although this perennial herb gets woody with age, it tends to die back in New York winters, and the new growth each spring is tender and flavorful. Common sage is the hardiest variety.

Salad burnet. Cooler than a cucumber, perennial salad burnet has an intense cucumber flavor in its attractive, serrated leaves. The more you use it, the more tender its leaves.

Summer savory. Less potent than the hardy winter savory, this mint relative has the ability to intensify the flavors of the foods it seasons. A classic pairing is summer savory and beans.

Thyme. One of the most versatile herbs, thyme has a subtle flavor like a cross between mint and oregano. It is at home in salad dressing as well as on grilled meat or fish.

Container Vegetable Gardening

Some edibles, such as corn and pumpkins, would be impractical in a pot, but you'd be surprised by how much food you can grow in a small space.

Many edibles grow happily in pots.

CONTAINERS

When you're choosing a container for edible plants, make sure it's not made of material that could leach toxic substances into the soil. Old tires and pressure-treated wood, for example, are not good options. In addition, the container must contain drainage holes; few plants like to sit in mud. Another consideration is the pot's color. Darker pots heat up faster and retain heat longer, particularly if you're container gardening on blacktop or in a hotter climate; therefore, the soil will dry out faster and the roots can be burned. If you live in an area with cool or overcast weather, however, darker pots could offer the additional warmth required by heat-lovers such as eggplants and melons.

SOIL

Use a quality, loose potting mix; don't use regular garden soil. Potting mixes comprising ground bark, peat, and vermiculite are able to hold water long enough for the plant roots to absorb it, but still allow excess water to drain through. Regular garden soil has a tendency to compact, preventing good drainage and air flow.

WATER

Containerized plants demand more and more water as they grow larger and the root mass expands. Check your plants daily by poking your finger into the soil; if the soil feels dry a couple of inches down, it's time to add water.

FEEDING

Because you are not building up the soil in the container the way you would by amending garden soil, you'll need to add some nutrients to the pot. You could add a small amount of compost to the mix (1 part compost to 2 parts potting mix) to enrich the mix, but your plants will need more. Luckily, many new organic fertilizers work faster than soil amendments. You can either incorporate a time-release fertilizer at planting or feed the plant every week or two with a water-soluble fertilizer as you water.

WIND

Something container gardeners often overlook is the effect of wind. Although some air circulation is good for plants, frequent gusts, which you might get on a balcony or rooftop, will tip over containers of tall plants such as tomatoes, trellised vines, and fruit trees. Wind can also shred leaves and knock fruits off branches. Provide some type of wind block in particularly windy locations. To help ensure that your pots don't tip over, make sure containers are wide and heavy enough to balance the weight of the top growth of your plants. You may need to anchor pots with bricks or heavy stones.

WHAT TO GROW

Space is not always an issue in edible gardening, because you can grow virtually any vegetable, herb, or fruit in a container.

Fruits. Many berry plants are particularly adaptable to containers, including strawberries, blueberries, currants, and gooseberries. They need large pots at least 2 feet wide and deep to balance their mature size. Choose a variety that is hardy to at least two zones cooler than your area if you will be leaving a container-grown fruit outside for the winter. Containers offer less soil to insulate roots than the plants would receive in the ground, and that extra measure of hardiness should offset the lack of insulation. The bigger the pot, the better the insulation.

Full-size tree fruits would not be practical in a pot, but some columnar fruit trees are bred for container growing. They are short and narrow and won't yield like a spreading tree, but you'll get more than enough fruit to snack on, and they are attractive on a patio, too.

Herbs. All but the tallest herbs do well in pots. Forsake the horseradish and go for bushy herbs such as oregano, mint, basil, and sage. You can bring annual herbs indoors for the winter, and most perennial herbs will be hardy enough to overwinter outside down to zone 5 or 6, but they can also be brought indoors to enjoy.

Vegetables. Vegetables in containers will require the same care you would give them in the ground, plus a lot more water. Dwarf or bush varieties are the best choices for containers, but if you want to plant a favorite variety, go ahead and give it a try. Fast-growing plants such as lettuce, beets, radishes, and leafy greens can be succession planted throughout the summer. Long-producing vegetables such as tomatoes, peppers, and zucchini pay off nicely all season long.

Bare-root planting

Many plants can be purchased and planted while they are dormant, with no soil around their roots. This is called bare-root planting, and it's a great way to buy plants inexpensively. Fruit trees, strawberries, and perennial vegetables such as asparagus, rhubarb, and horseradish are often sold as bare-root plants. Bare-root trees and shrubs can look like lifeless twigs and perennial plants can resemble a mop of roots, but they are very much alive. They can't survive for long in this state, however, so it's very important that you get them in the ground as soon as you can. If their permanent spot is not ready for planting, plant them in a temporary holding bed and plan to move them later.

STEPS:

1. As soon as your plant arrives, remove it from its package and take off any packing material, such as shredded newspaper or peat moss.
2. Check to make sure the plant is healthy. The roots should feel fleshy, not dry, and they should smell fresh and earthy, with no rotting odor. Carefully prune away any damaged roots or shoots.
3. To ensure that the roots are well hydrated before planting, soak them in water for an hour or two. Don't leave them in water longer than overnight or they can start to rot.
4. Dig a hole wide enough to allow for the roots to fan out. Then make a mound of soil in the center of the hole, high enough so the crown of the plant will sit at or above the soil line. It is very important that the crown is not buried.
5. Center the plant on the top of the mound and spread the roots around and down.
6. Fill the hole halfway with soil, and then water well. After the water has drained, finish filling the hole and water again.
7. Apply a layer of mulch.
8. Water the plant every 2 or 3 days, whenever the soil seems dry below the surface, until you see leaves sprouting. Then you can move to a watering schedule of 1 or 2 inches per week.

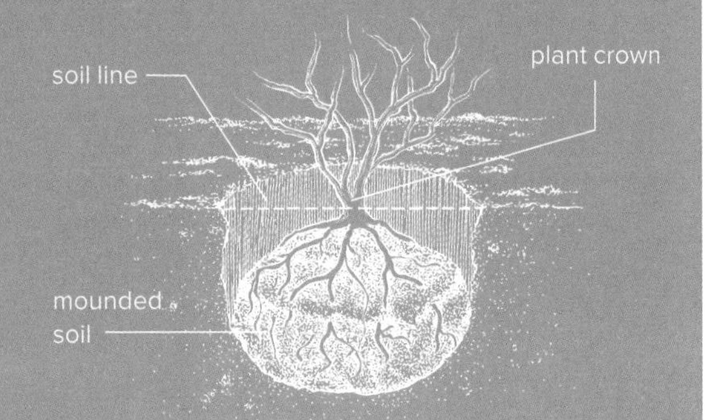

▲ When planting bare-root plants, mounding the soil in the center of the planting hole will help you adjust the plant crown to just above the soil line, while letting the roots fan out and down.

JUNE
Growing Up

It's officially summer. Just about everything has been planted and much has been harvested. June requires hustle, because it's time to start realizing all the gardening plans you made earlier. If you planned to swap out the spinach for tomato plants, you can harvest the spinach and plop a tomato seedling in its place. And then, what the heck, seed a little more spinach nearby—the shade from the tomato plant might let you grow both vegetables at the same time. In June, you'll plant long-season vegetables that take up a lot of space. If you're wondering where to squeeze them in, take a deep breath and read on. There's always more space if you grow your plants vertically, and they might even thank you for it.

◀ June is the month for strawberries, but be sure to protect them from pests and changeable summer weather.

TO DO THIS MONTH

PLAN

- Make notes in your garden journal for next year
- Scour nurseries for late-season deals

PREPARE AND MAINTAIN

- Thin seedlings of beets, carrots, and salad greens
- Set up trellises for vining crops
- Place cutworm collars on new transplants
- Remove blossoms from new straw-berry plants

Zones 5, 6, and 7

- Hill potatoes
- Look for cabbage worms, corn borers, slugs, squash vine borers, and whiteflies
- Prune tomato suckers
- Pinch back basil

Zone 3 Zone 4 Zone 5 Zone 6 Zone 7

SOW AND PLANT

Everyone

- Continue succession planting
- Make sure the garden is getting at least 1 inch of water a week

Zones 3 and 4

- Direct sow: beans, corn, cucumbers, melons, and squash
- Transplant outdoors: basil, celery, chicory, cucumbers, eggplants, endive, kohlrabi, melons, mint, okra, peppers, pumpkins, squash, thyme, and tomatoes
- Continue succession planting: beets, broccoli raab, carrots, green onions, kale, and lettuce

Zones 5, 6, and 7

- Plant something extra to donate to a food kitchen
- Direct sow: beans, corn, cucumbers, melons, and squash

HARVESTING NOW

- Arugula
- Asparagus
- Beets
- Blueberries
- Broccoli
- Broccoli raab
- Cabbage
- Chard
- Corn
- Currants
- Fava beans
- Garlic scapes
- Green onions
- Herbs
- Kale
- Kohlrabi
- Lettuce
- Radishes
- Rhubarb
- Spinach
- Strawberries
- Zucchini blossoms

Supporting Your Plants

No matter how much space you have, sometimes the best way to grow vegetables is up. Supporting plants off the ground lifts them away from ground insects, roving animals, and soil-borne diseases. It keeps the vegetables clean and the plant's branches untangled. Hanging vegetables can grow large without hindrance, and they will be easier to find and harvest. And, of course, vertical growing saves space. Vegetable garden real estate is valuable.

Two types of plants can be grown on supports: the type that cling on their own, such as peas and squash, and those that vine but don't cling, such as tomatoes. The type of plant you're dealing with will help determine what type of support you use.

TRELLISES

Long, sprawling vines such as squash will take up as much space as they can get away with on the ground. They are likely suspects for trellising. Tomatoes and pole beans will flop and tangle, rather than sprawl, and they can also benefit from some type of support. You will need to do a little training when the plants are small to get them leaning in the direction of the trellis, but once they grab hold they will do the rest of the work.

For lightweight vines, you can make your own grid by tying and weaving biodegradable twine made from fibers such as jute, sisal, or cotton between the supports. At the end of the growing season, you can toss the whole thing into the compost—which is handy, because some vines can be difficult to pry off a trellis. Large plants require a stronger trellis, and even then, trellising works best for varieties that produce fruits in the range of 2 to 4 pounds. Large, heavy pumpkins, for example, will not only snap a trellis, but they can also snap their own stems and fall to the ground, unripe.

A trellis shape can be linear, decorative, teepee, or A-frame. The main supports should be long enough to drive at least 1 foot into the ground. A basic trellis could be two 6-foot wooden stakes spaced 5 or 6 feet apart, with netting or wire fencing stretched between them. You can attach the netting with ties, nails, or long staples. A center bar across the top will make the trellis a lot more strong and stable to support the vines and their fruit.

Trellises angled to maximize sun exposure can also be used for tomato plants in espalier fashion, with the branches fanned out along the netting. As the plants grow, you can stretch out the branches and tie them to the trellis to create a flattened, two-dimensional

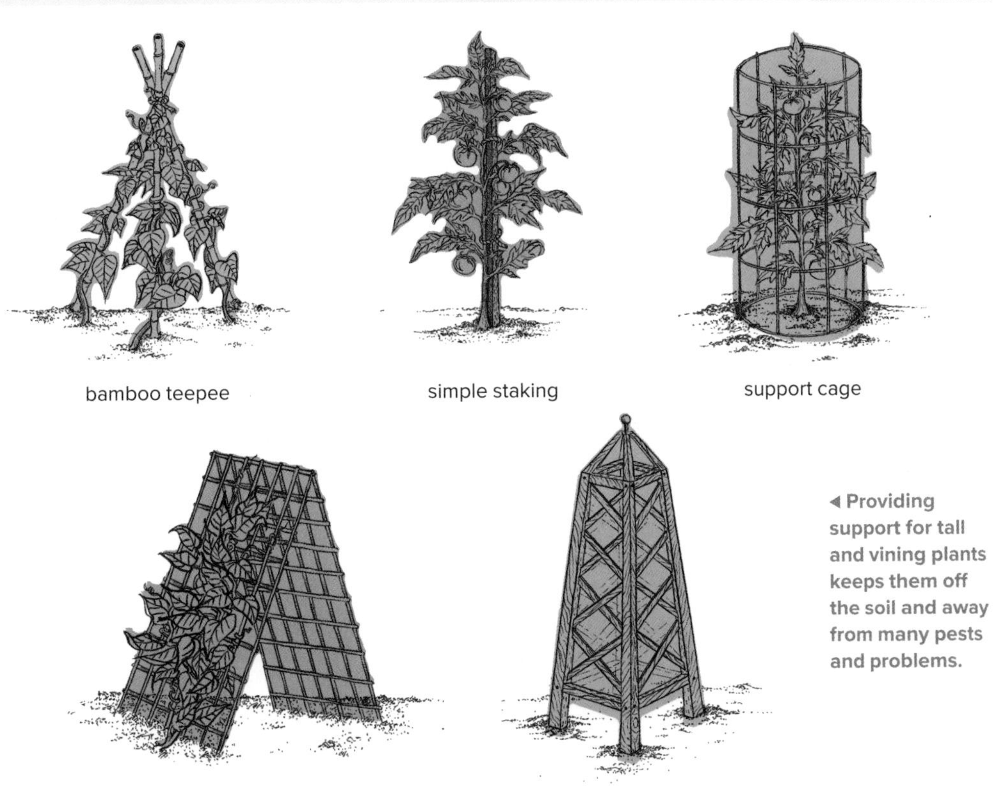

◀ Providing support for tall and vining plants keeps them off the soil and away from many pests and problems.

effect that not only supports the plants but allows the sun to reach all the fruits.

Whatever plants you are trellising, set them within a few inches of the trellis so the plants don't have to work hard to find it, and use the suggested plant-spacing guidelines. It's fine to mix varieties on one trellis, as long as you don't overload the structure.

STAKES

Plants that grow tall and become top-heavy with fruit may require only a straight stake to keep them upright. (It's best to position the stake at planting time; if you wait until after the plant is established, driving in the stake can injure the plant's spreading roots.) Tie the stem to the stake, and then add ties higher and higher as the plant grows—but don't tie it so tightly that the stem rubs against the stake. Give the stem a little room to continue filling out and to sway in the breeze, which makes it stronger. Use a tie that stretches, or make a figure 8 when you're tying by wrapping twine around the stake and then crossing it over itself, before wrapping it around the stem. Bamboo stakes are fine for shorter plants with lightweight fruits, such as peppers. Heavier plants or fruits require 1-by-1-inch hardwood stakes, sometimes called tomato stakes. These are sturdier and thus less likely to snap under the weight.

CAGES

Cone-shaped wire cages were traditionally used for tomato plants, but most are too small

MY A-FRAME TRELLIS

For heavy vines, I construct an A-frame using four 6-foot fencing stakes and 4-foot-wide plastic netting. I sink two of the stakes about 1 foot into the ground, approximately 3 feet apart, and lean them toward each other until the tops meet, and then tie or screw the stakes together. I do the same with the other pair of stakes 4 feet down the row. For extra netting support, I run a metal pole across the tops of the stakes, resting it in the joints where the stakes meet. Then I'm ready to drape the netting over the whole thing. I can plant and harvest along both sides of the trellis as well as underneath it. While the vines are filling in, I make use of the space under the A-frame to plant shade-loving greens such as lettuce and spinach.

for vining tomatoes. They do make quick, convenient supports for other bushy plants, however. Newer cages are much larger and sturdier than the older types we used to find at nurseries. You'll find cages up to 5 feet tall, with large holes to harvest through. Look for cages that fold flat for storage; they can be a bit more expensive, but they will last virtually forever. If you are industrious, you can build your own cages out of concrete reinforcing mesh, which is the perfect strength and size but difficult to work with. You'll need strong hands and heavy-duty wire-cutters. Look for a roll with a 4- to 6-inch mesh; you can cut a 5-foot length and roll it into a cage, securing the ends with wire or ties. Wire mesh lasts many growing seasons, so your initial effort will pay off in the long run.

Growing Strawberries

June is the traditional strawberry month. June-bearing strawberries are a fleeting treat, but one we always eagerly anticipate. New York gardeners have three choices of strawberry plants.

June-bearing strawberries set their whole crop at once and ripen within a few weeks in June. You can get a longer harvest season by planting a mix of early-season, midseason, and late-season varieties.

Everbearing strawberries do not produce continuously, but they set fruit several times throughout the season. The largest crop appears in the spring, followed by smaller crops in summer and fall.

Day-neutral strawberries flower and fruit throughout the season, producing small and consistent yields.

Day-neutral and everbearing strawberries have overtaken June-bearing strawberries in popularity for home gardeners. They don't produce as many runners or side shoots as June-bearers, so they require much less space and maintenance. However, if you want to preserve strawberries and want a large crop all at once, June-bearers are the way to go. If you want a steady supply of strawberries all summer, try one of the others.

PLANTING AND CARING FOR STRAWBERRIES

Strawberry plants are usually sold as bare-root plants. They may not look like much, but they will perk up quickly after they're in the ground. You can plant strawberries any time after the soil has warmed and dried.

Choose a site that gets at least 6 hours of sun per day. Amend the planting area with 2 to 4 inches of compost. Strawberries need

Plant strawberries where they will get at least 6 hours of sunlight a day.

a rich, well-draining soil. If your soil is poor, add some granular organic fertilizer a few weeks before planting.

Strawberries' long, thick roots attach at a center disk called the crown. Plant strawberries so that the roots are fanned out and the crown is just above soil level. Do not cover the crown with soil or it will rot and the plant will die.

The plants will show signs of growth within a couple of weeks. In a month or two, they will start to send out runners—new shoots that will eventually become separate plants. This is where the care of the different types of plants diverges.

June-bearing strawberries are easiest to cultivate if you space new plants about 20 to 30 inches apart in rows spaced every 3 to 4 feet. As the runners are sent out, they will root between the existing plants, form daughter plants, and create a mat of strawberries. Don't allow rows to spread more than 2 feet or you will not be able to walk between them to weed and harvest.

Remove all the flowers for the first year. This is tough to do, I know, but it will pay off for years to come. Letting the plants put their initial energy into developing roots and leaves will result in strong plants, lots of runners, and eventually lots of sweet strawberries.

The real work of June-bearing strawberries comes in renovating the beds each season. After harvesting the fruits, mow or trim the whole patch to about 1 inch of growth, but don't mow down to the crowns. Rake off and remove the mowed leaves. Remove the plants on the outside of the rows, so that each row is only 6 to 12 inches wide. You can dig them out or till them in. Then thin the remaining

PLANTS THAT CAN BE TRELLISED

Some plants simply must be trellised, and for others, trellising just makes good sense. These are my tips for matching the vegetable to the trellis.

Beans. Bush beans can support themselves, but pole beans need something to cling to, and it had better be tall. Many varieties can grow 12 feet tall and beyond. Of course, it would be difficult to harvest beans 6 feet over your head, so pole beans are traditionally grown on teepee trellises and allowed to circle around. Make the teepee using three 6-foot bamboo stakes of about ½ inch in diameter. To give the vines plenty of area to grab hold, weave twine in a cross pattern around the stakes. Plant five or six seeds at each stake.

Cucumbers. Cucumbers are prone to powdery mildew, and trellising can help prevent this by exposing the vines to more sun and air. The A-frame trellis would be a good choice for these heavy plants. Plant cukes on the side opposite the afternoon sun and they will scramble toward it.

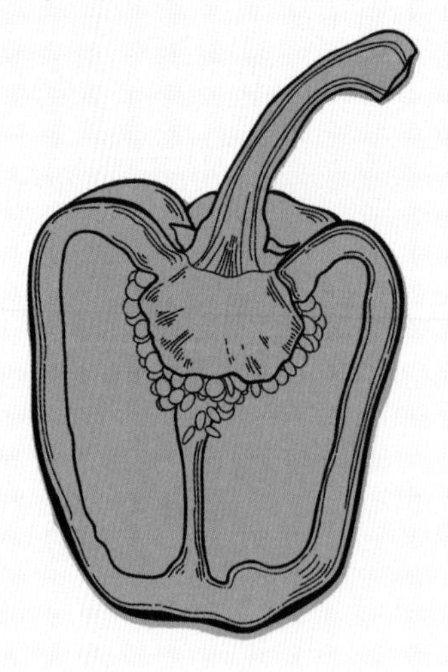

Eggplants and peppers. These tomato cousins need some support as they mature and become top-heavy with fruits. They don't vine or sprawl, but stems and branches will flop and possibly snap from the weight. A traditional tomato cage should suffice, or you could use the staking method and tie them to it a couple of times as they grow.

Melons. Because melons are heavy vines, the A-frame trellis is the best choice. Some melons, muskmelons (cantaloupes) in particular, slip or separate from the vine when they are ripe. You can create a sling around the fruits to prevent them from crashing to the ground. Tie strips of old T-shirts or tights—anything with some stretch—around the fruits and secure them to the trellis.

Peas. Some pea varieties are taller than others, but they all benefit from trellising, which helps keep them dry and pest-free. Peas are not heavy vines, and even mesh or twine stretched between two poles or an A-frame will be sufficient. Set up pea trellises early so you'll be ready to plant early in the season.

Squash and pumpkins. Squash plants are probably the heaviest vines, so build the sturdiest trellis you can and plant only one or two plants every 4 feet. I find large pumpkins too heavy to hang onto the vine when trellised, so I allow them to grow on the ground. If you are short of space, you can always cradle the fruits in a sling like the one described for melons.

plants to a 4- to 6-inch spacing. Top dress with compost and a slow-release organic fertilizer and water weekly. The plants will revive and start producing new runners for next year's strawberries.

Day-neutral and everbearing plants produce best if the few runners they produce are removed from the plants, because much of the plant's energy will go into growing the crown and flower stalks instead of the berries. Plant them about 1 foot apart and mulch between plants to retain moisture, suppress weeds, and keep the fruit off the ground. Fruits resting on the ground are more susceptible to rotting and to soil-borne diseases. Remove the first flush of flowers from new plants to give them some extra time to develop roots and vigor. After that, allow them to flower freely.

All strawberry plants need at least 1 inch of water per week, or more in extremely hot weather. Strawberries need some supplemental feeding starting in their second year. Wait until they have produced their first crop and then feed them a balanced fertilizer. Don't be tempted to overdo it, or your plants will produce lots of lush growth and no flowers or berries.

Although strawberry plants are hardy, they can be damaged by spring frosts or from heaving out of the soil with repeated freezing and thawing during winter. To prevent this, cover the area with a layer of straw (and that's how they got their name), shredded leaves, or a row cover. Be sure to remove the cover as the ground warms in spring, but be prepared to cover them again if a frost threatens. Then wait for a whole new crop of strawberries to start setting and ripening.

Tomatoes. Tomato plants grow quickly, and continually having to tie them to a support as they grow can be a chore. After they start setting fruit, you may find it difficult to lift the stems. To make this easier, you can set out tomato seedlings in cages. As the plants grow, they intertwine within the cage, which hopefully holds the plants up. Sometimes tomato plants don't like to cooperate, so some initial training may be necessary.

Traditional cone-shaped tomato cages are pretty useless for supporting most tomato plants. They are too short, and the openings are too narrow to reach through. The plants will eventually grow above and through the support and flop on the ground. Look for a cage that is at least 5 feet tall, with access holes about 4 inches wide. I like the Texas Tomato cages, because they're tall, sturdy, and fold flat for storage.

How to plant a tomato

It's hard to go wrong when you're planting a tomato, but you can do a few things to ensure a hardier plant.

Don't rush it. Tomatoes do not like the cold, so wait until both the air and soil temperatures stay above 50°F.

Make sure you harden off seedlings. The shock of going from the nursery to chilly nighttime temperatures or even bright, hot sun will set back blooming time.

Plant seedlings deep. Tomatoes can grow roots all along their lower stems. Those little bumps you see on the stems are waiting to grow adventitious roots, and burying part of the stem is all the encouragement they need. More roots mean more water and nutrients going into the plant.

To get those roots growing, you can either dig a deep hole and plop the plant in, or make a long, shallow trench and lay in the plant sideways. Either method is fine, as long as you leave only the top one or two sets of leaves above ground. You don't even have to remove the other leaves when you are planting. Don't worry if you go the trench route and the plant is a little lopsided; it will eventually find the sun and straighten out. This method works especially well if seedlings have gotten leggy or spindly. And don't forget to put your stakes or cages in place at planting time; you don't want to damage those extra roots later.

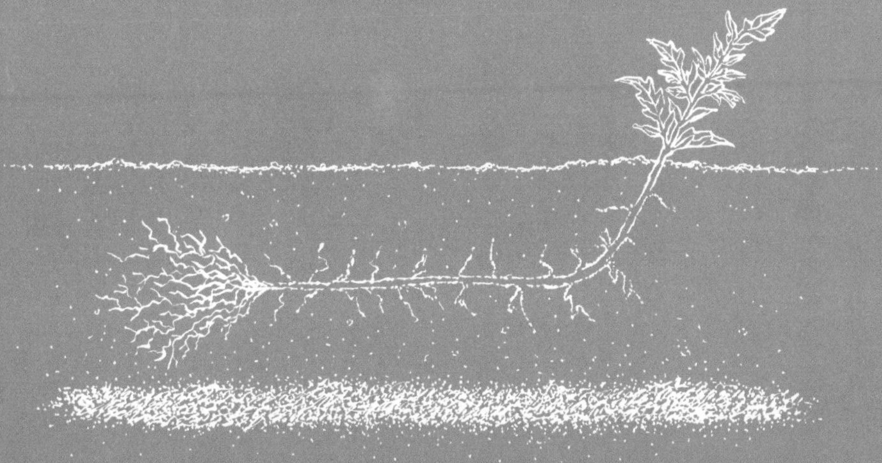

◀ Burying the tomato plant's stem will produce more roots along it and lead to a stronger tomato plant.

TOMATO SPEAK

- **Determinate:** Sometimes called bush tomatoes, these plants grow to a fixed size and set and ripen the bulk of their fruit within a 2-week span. Many paste tomatoes are determinate.
- **Indeterminate:** These vining plants continue to grow and set fruits throughout the season. Most tomatoes are indeterminate.
- **Heirloom:** Heirloom tomatoes are open-pollinated plants whose seed has been saved and handed down for generations. There is a great deal of variety in heirloom tomatoes.
- **Hybrid:** Hybrid tomatoes were bred to have some type of feature, such as disease resistance or uniform color.
- **Cherry:** These small, round, sweet tomatoes tend to grow in clusters and are often devoured by gardeners before they make it into the house. Other minis, grape tomatoes, are oblong in shape. They are crosses of paste tomatoes and tend to have less water than cherry tomatoes, and they are a little sweeter, although a lot depends on the variety and growing conditions
- **Slicer:** These tomatoes are regular, round, and juicy. The larger slicers, generally weighing a pound or more, are called beefsteaks.
- **Paste:** Also known as Roma or plum tomatoes, these are traditionally used for sauce and drying. They have a lower water content and a more concentrated flavor.

JULY
Problems Happen

July's weather can be challenging: muggy with frequent harsh storms, punctuated by periods of drought. This seems to be a big attraction for insects, however, as they move in seemingly overnight and set up camp in the garden. It's also perfect weather for perking up every dormant fungus spore in the soil and every exposed weed seed. Plant diseases can sneak into the garden, especially when the weather is damp or humid—ideal conditions for many types of diseases. All of this makes for stressful conditions for your plants. No matter how well you have planned and how many precautions you take, your garden will experience some problems—we are dealing with nature, after all. July presents less-than-ideal gardening conditions, but we have to learn to work with them. Stay vigilant.

◀ July is a month of plenty—enjoy the bounty, including fresh greens from the garden.

TO DO THIS MONTH

PLAN

- Make notes of planting dates and weather influences in your garden journal
- Make plans for your fall garden

PREPARE AND MAINTAIN

- Side dress with compost or fertilize
- Look for cabbage worms, corn borers, slugs, squash vine borers, and whiteflies
- Hill potatoes and dig out a few new potatoes for dinner
- Prune tomato suckers
- Thin and eat beet, carrot, and salad greens
- Pinch back basil
- Keep an eye out for cutworms, flea beetles, and leaf miners

Zone 3 Zone 4 Zone 5 Zone 6 Zone 7

SOW AND PLANT

Everyone

- Fill in vacant spots in the garden
- Sow indoors: Asian greens, broccoli, Brussels sprouts, cabbage, cauliflower, kale, and Swiss chard

Zones 3 and 4

- Direct sow: beets, carrots, green onions, kale, lettuce, and radishes
- Continue succession planting: beans, beets, cucumbers, and lettuce

Zones 5, 6, and 7

- Make final succession plantings of bush beans

HARVESTING NOW

- Basil
- Beans
- Beets
- Blueberries
- Cherries
- Chicory
- Cucumbers
- Early greens and herbs
- Eggplant
- Endive
- Garlic
- Lettuce
- Melons
- Onions
- Peaches
- Peas
- Peppers
- Potatoes
- Raspberries
- Rhubarb
- Green onions
- Summer squash
- Tomatoes

Conserving Water in the Garden

In New York, water is often in short supply during the summer. When the lawns turn brown, you can bet water restrictions are at least being considered by your town. To avoid losing your vegetables during dry spells, follow some guidelines.

Water appropriately. Plants need water at regular intervals; they don't have reservoirs for storing up extra water until they need it. Try to give your vegetable garden at least 1 inch of water every week. Check to make sure that the soil gets wet at least 6 inches below the surface. A slow soaking with drip irrigation, a soaker hose, or a fine mist sprinkler can accomplish this.

Add mulch. We position our vegetable gardens to take advantage of as much sun as possible. That's great for the plants, but it can bake the soil. A 3- to 4-inch layer of organic mulch, such as straw or shredded leaves, will shade and cool the soil and conserve moisture. Plastic mulches are also beneficial in the vegetable garden, but they tend to warm up faster than organic mulches, and you may need to water more often.

Don't water at midday. Water when it is hot and sunny, and you waste a significant percentage of the water to evaporation before it even reaches the ground. Morning watering is best: it hydrates the plants and makes them ready to face the day ahead. Early evening watering is better than midday watering, but watering in the evening can leave plants damp in the cooler night air, which can promote fungal diseases.

Improve poor soil. Water runs right through sandy soil and runs right off of clay. To keep the water where plant roots can access it, regularly work compost and other organic materials into the soil. These amendments act as sponges to hold water in.

RAIN BARRELS

Be prepared for drought by saving water when it is available. The simplest way to do that is to install a rain barrel or two. Collecting water in rain barrels is rapidly regaining popularity with gardeners as a way to ensure that water is available for plants even when there's no rain in sight.

Consider that it takes about ½ gallon of water per square foot to get to that required 1 inch of water per week. That means that a 10-by-10-foot garden would need 50 gallons of

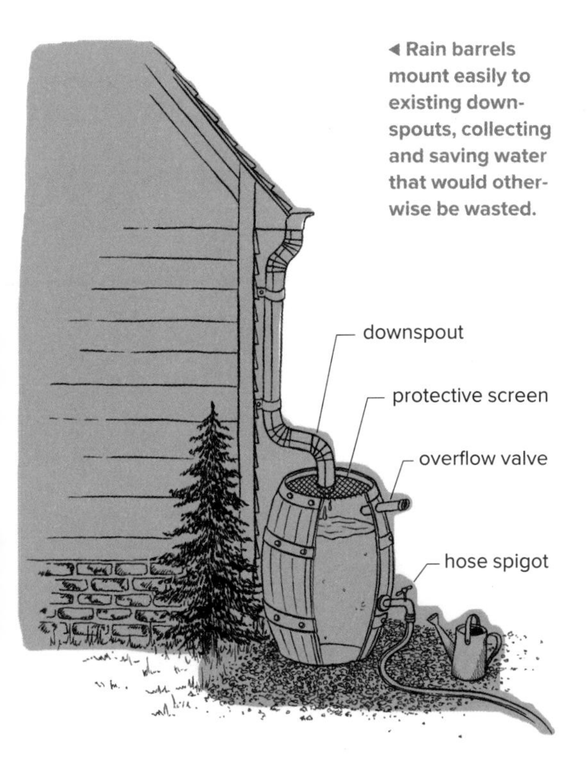

◀ Rain barrels mount easily to existing downspouts, collecting and saving water that would otherwise be wasted.

water each week, or about 600 gallons during the three summer months. That's water you have to pay for or water from the well you rely on for drinking. By attaching rain barrels to your gutter downspouts, you can store thousands of gallons of water during the times when the rain does fall. One inch of rain on a 1000-square-foot roof can provide 600 gallons of water that is soft and virtually chemical-free. And in New York, you can be assured that you'll be getting a lot more than 1 inch of rainfall every year.

Rain barrels are sold at garden centers, hardware stores, and online, with styles for every taste; most require minimal assembly and installation. Most rain barrels will store 55 to 80 gallons and can easily be filled during a single rain storm. Consider adding two or more barrels to store even more water. You withdraw water from a spigot near the bottom of the barrel, which can be connected to a garden hose.

To make the most of a rain barrel, make note of a few guidelines.

- Choose a barrel made from food-grade plastic that offers ultraviolet stability (UV stable).
- Be sure to place your rain barrel on level ground.
- Make sure the cover fits tightly to prevent kids and animals from falling in.
- All barrel openings should be fitted with screening so that the water does not become a mosquito breeding ground.
- Use the water or disconnect the downspout when the barrel is full. Overflow can seep into basements and damage foundations.
- Disconnect the rain barrel, empty it, and store it upside down for the winter. Repeated freezing and thawing will cause cracking.

Tomato Problems

No plant is more prone to problems than the ever-popular tomato, especially in the humid Northeast. We simply have to assume that something will affect our precious plants at some point during the growing season. Very often it's a cultural problem: too hot, too cold, too wet. Disease can also affect tomato plants. Fortunately, insects aren't much of a problem for tomatoes, except for one: the tomato hornworm.

You can check for a few causes before you look for a cure. Most tomato problems first start showing symptoms on the leaves—usually as spotting. If you can stay alert and watch for symptoms of problems early, most of them can

Inconsistent watering can result in cracked tomatoes. They are fine to eat, but do so as soon as possible.

be controlled long enough to keep the plant growing throughout the season. A few diseases, however, have no controls, so pulling out and destroying the plant as soon as you notice the problem is your best bet at saving the remaining tomato plants.

Here's a roundup of common tomato problems in the Northeast, starting with diseases.

Early blight. Also known as Alternaria leaf spot, this fungal disease causes dark spots on the leaves, starting with the older leaves. If left unchecked, it will also affect the fruits. Remove affected leaves and water the plant only at soil level. Fungicides containing sulfur can slow the spread.

Fusarium wilt. This fungal disease clogs the plant's vascular system and blocks the flow of water and nutrients. Older leaves start to turn yellow and droop. There's not much you can do after plants are infected. Next year, choose a variety that is resistant to Fusarium wilt.

Late blight. This fast-spreading disease can quickly kill tomato and potato plants. On tomatoes, leaves and then fruits develop irregularly shaped gray spots and a greasy appearance. A ring of white mold may appear around the spots and stems may blacken. Because there is no cure, destroy infected plants—do not compost them. Notify your local cooperative extension office, because late blight can spread for hundreds of miles and wipe out commercial fields. If you hear that late blight has been verified in your area, but your plants are not yet affected, spray them with a copper fungicide or a botanical fungicide called Serenade, which seems to provide some resistance.

Septoria leaf spot. One of the most common fungal diseases appears on the leaves as white or gray spots with darker borders. As with early blight, sulfur sprays can slow the spread of the fungus.

Verticillium wilt. Yet another fungal disease, Verticillium wilt can cause the plant to wilt during the day, with recovery at night; or the leaves may turn yellow and dry up. There is no cure, and the spores persist in the soil for years, so the best advice is to choose resistant varieties.

Tomatoes aren't bothered by many insects in New York, but the major insect pests, tomato hornworms, are so obnoxious that they will devour whole plants in a short time. The larvae of the hawk or sphinx moth, also known as hummingbird moth, are large, green caterpillars as big as your pinky finger. They are large enough to remove by hand, and you can also encourage predatory wasps or use *Bacillus thuringiensis* (Bt), a biological insecticide used to control caterpillars. The bacterium affects only moth and butterfly larvae, so other insects are not harmed. (Keep in mind that Bt is a pesticide and should be used only when pest populations are out of control, and always follow label instructions.) Aphids and whiteflies can also infest tomatoes if air circulation is poor. You can blast them off the plant with a strong spray of water.

Problems caused by less than optimal growing conditions, such as too much or too little water, sunshine, or heat, can be the most frustrating to deal with. We can't control prolonged rains or heat waves, but we will need to compensate for them by turning off the irrigation system in wet weather or giving the plants extra water when they're parched. Pay attention and stay on top of the situation, because most cultural problems don't become apparent until the fruits start to mature, when it's too late to correct them.

Catfacing. This term refers to distorted, misshapen fruits. The most likely culprit is a cold spell that occurs while the plant is in bloom. Don't be tempted to set your seedlings out too early.

Blossom end rot. Dark, sunken blotches at the blossom end of the fruit are caused by a combination of conditions. Excessive hot or cold weather while the flowers are setting and fluctuation in the amount of watering combine to limit the amount of calcium distributed throughout the plant. The problem starts as the fruit is forming but doesn't show up until it's too late to stop it. Again, don't rush to get your seedlings out. Be consistent about watering, and make sure you fertilize the plants appropriately.

Cracking. Inconsistent watering is the problem here. If your tomatoes go from the dry side to being heavily watered, the pulp inside the tomato plumps up faster than the outside skin can stretch. The fruit will crack open, but they're still good for eating. Use them as soon as possible, however, or the cracks could invite other problems, such as mold.

Green shoulders. Sometimes tomatoes never seem to ripen all the way around. Green at the stem end could be the result of hot temperatures—or it might just be a characteristic of the tomato variety; several heirlooms naturally have green shoulders.

PLANT MATURITY

PLANTS THAT MATURE IN 30 DAYS

- Arugula
- Beets
- Mustard
- Radishes
- Spinach

PLANTS THAT MATURE IN 60 DAYS

- Basil*
- Beans (bush)*
- Broccoli
- Carrots (baby)
- Cilantro
- Collards
- Green onions
- Kale
- Kohlrabi
- Leeks
- Lettuce
- Swiss chard
- Turnips

PLANTS THAT MATURE IN 90 DAYS

- Brussels sprouts
- Cabbage
- Carrots
- Cauliflower
- Fava beans
- Parsnips
- Peas
- Rutabaga

**Plants killed by frost*

Sunscald. Sun is good for tomatoes, to a degree. When the fruits are exposed to excessive sun, however, they can actually burn. This often happens when another condition has caused the loss of too many leaves. Sunscald looks like light, papery patches on the green fruits.

Planning for Fall

I know. You've just gotten things under control and you're really starting to enjoy your garden. Why do you need to prepare for fall? It's months away. But it's really not too soon to prepare—at least not if you want your garden to continue on past the peak of summer. You can sow more vegetables now to mature and harvest in the fall.

If you don't plan on protecting your garden crops after frost, you'll need to check for your first fall frost date and determine how long it takes the vegetables to mature; then count backward to determine when to plant. Most plants require at least a couple months to mature, which means that planting in July will give you a harvest in September. (And if you get your cold frame up, you could still be dining from the garden during the holidays.)

Add some compost and organic fertilizer to replenish the soil, and get planting. This is also the time to resume succession planting of cool-season vegetables. If you don't want to plant a fall vegetable garden, July is a great time to start seeds for a cover crop.

IS IT RIPE YET?

Nothing tests your patience like a vegetable garden, but there is no sense in rushing things. As the old saying goes, "It takes as long as it takes." Picking vegetables and fruits before they are fully ripe will only disappoint you. On the other hand, you do not want to wait too long so that cabbages split down the center and beans get stringy. Harvesting our crops at their peak is more art than science, but you will get the hang of it. Watch for definite signals, such as a fruit reaching full color and size, having a sweet aroma and a crisp or tender texture, and a feeling of weight. You'll need all five senses, a little good judgment, and a few tips.

Asparagus should be picked while pencil thin, when the stalks snap off easily.

Beans (snap) should look smooth and firm and should snap off the vine. Don't wait until you see seeds bulging through the pods.

Beets, radishes, and carrots usually poke their shoulders above the soil when mature, but you can gently poke around under the soil to see how they're doing.

Berries take their time ripening, dramatically changing color and getting larger, softer, and more sweetly scented. But you'll still need to taste some to know for sure.

Broccoli is ready when the head looks and feels firm.

Cabbage heads should feel solid when squeezed.

Cauliflower should be harvested while the curds (the heads) are still smooth, not grainy.

Corn is easy. When the silks turn dry and brown, open a husk to see if the kernels exude a milky substance when pricked.

Cucumbers and summer squash should be firm, smooth, glossy, and tender enough to poke a fingernail through. Don't wait for them to get large. Oversized fruits become tough, bitter, and seedy. Harvesting while they are young and tender will encourage the plants to keep setting more fruit.

Greens can be harvested when young and tender or allowed to reach their mature size and crispness.

Melons will change color and the fruits will smell sweet. Muskmelons (cantaloupes) will slip right off the vine.

Pea ripeness can be judged by how full the pod feels, but tasting truly tests peas for readiness.

Potatoes offer two ripeness choices. New potatoes can be scavenged when the plant flowers. For full-size potatoes, hang on until the plants begin to dry out and turn brown.

Pumpkins and winter squash will fully change color and the vines will start turning brown, with no new growth.

Tomatoes send out three cues: ripe fruits should be fully colored, aromatic, and soft to the touch.

Watermelon is an enigma. Thumping for ripeness can take a lifetime to master. Instead, check for when the white spot on the bottom of the melon changes to a rich yellow color.

The best time to harvest most vegetables is in the morning. Plants are still filled with water and will not wilt before you can get them indoors. Staying on top of harvesting while trying to pick only what you can eat fairly quickly can be a challenge. Greens and root vegetables will store well if they're kept slightly damp and in the refrigerator. Juicier fruits and vegetables stay freshest when stored on a cool counter; refrigeration diminishes their flavor.

Learning to identify pest problems

Tomatoes are not the only vegetables that attract unsavory characters to your garden. Although you know that not all insects are bad guys, some are definitely up to no good. You will never control them all, but you can take steps to keep them in check.

The first thing to do is be on guard. Keep a watchful eye on the garden and catch problems when they are small. When you spot a suspect insect, try to identify what it is. Is it a bad guy? Is it an insect that will be gone in a few days or weeks, or is it looking to invite its extended family for a feast?

Even if it's the latter, don't panic. Your garden is a balanced ecosystem with living soil, healthy plants, and beneficial insects that will prey on them, right? I hope so. Unfortunately, however, that's not quite enough. We are not the only creatures who find this region a welcoming place to live. In addition to the opportunistic insect, several diseases love to linger here. Sometimes you will need to take action to protect your plants. Instead of indiscriminately reaching for the first available spray can, good gardeners approach these problems methodically. Start by assessing the enemy by the damage it is causing.

THE WORST NEW YORK VEGETABLE GARDEN OFFENDERS

TYPE OF DAMAGE	THE USUAL SUSPECTS
Holes in leaves, chewed leaf edges	Chewing insects: larvae (cabbage looper, cabbage worm, cutworm, squash borer, corn borer, and hornworm), beetles (asparagus beetle, Colorado potato beetle, flea beetle, Mexican bean beetle, and striped cucumber beetle), slugs, and snails
Plant cut off at the base	Cutworms
Scarred, stippled leaves (and possibly sticky honeydew and black sooty mold)	Sucking pests: aphids, squash bugs, thrips, and whiteflies
Tunneling on root crops, plants wilt despite watering	Root-eating pests: cabbage maggot, carrot rust fly larvae, and squash vine borers
Tunnels within leaves	Leaf miners
White, powdery coating on leaves and stems	Powdery mildew and downy mildew
Dark spots on leaves or stems that may spread to fruits	Any of several fungal diseases

AUGUST
Abbondanza!

August is the month of plenty. Sometimes it's the month of too much. Don't let your fruits and vegetables go to waste. Preserving will make them available to enjoy well into winter, when the pickings are slim. Preserving your harvest doesn't have to be complicated and time-consuming. It can be as simple as tying up bunches of herbs to dry or finding a place where your stored potatoes will stay cool. With the harvest peaking, a gardener's job becomes more about replenishing the garden than taking from it. It seems only fair, after all. Start getting the soil ready for next season. Even though your plants are slowing down, don't hang up your trowel just yet. You still have plenty of gardening to do.

◀ **If your August garden gives you more than you can eat, pickling, drying, and canning can prevent waste and preserve the taste of summer for colder months.**

TO DO THIS MONTH

PLAN

- In your journal, make notes of planting dates and weather influences
- Order garlic for fall planting

PREPARE AND MAINTAIN

Zones 5, 6, and 7

- Water, water, water
- Remove flowers from tomatoes, winter squash, and pumpkins
  TIP ***Remove flowers so that existing fruits will ripen.***
- Harvest and cure mature onions and potatoes
- Prune water sprouts from fruit trees
- Prune raspberries and blackberries after fruiting
- Fertilize strawberries
- Harvest herbs to dry or freeze

Zone 3 Zone 4 Zone 5 Zone 6 Zone 7

SOW AND PLANT

Everyone

- Sow cover crops
- Direct sow: arugula, beets, carrots, kale, lettuce, and spinach
- Transplant outdoors: Asian greens, broccoli, Brussels sprouts, cabbage, cauliflower, kale, and Swiss chard

HARVESTING NOW

- Basil
- Beans
- Beets
- Blueberries
- Broccoli
- Cabbage
- Carrots
- Cauliflower
- Celery
- Chicory
- Corn
- Cucumbers
- Eggplant
- Endive
- Garlic
- Herbs
- Kale
- Leeks
- Lettuce
- Melons
- Nectarines
- Onions
- Pears
- Peppers
- Potatoes
- Radishes
- Raspberries
- Green onions
- Squash
- Strawberries
- Swiss chard
- Tomatoes

Preserving Your Harvest

Preserving vegetables and fruits can be a complicated process, or it can be as easy as letting them dry and popping them in a bag. It all depends on what you are preserving, how ambitious you are, and how much storage space you have. No one wants to see their bounty go to waste. If you make a plan for what you want to preserve and how you want to preserve it, you'll be prepared when all that lovely fresh produce starts piling up.

CANNING

Vegetables and fruits with a high water content, such as tomatoes, peaches, and pears, retain their flavors well when canned. Can them au natural or use them for sauces, jellies, jams, pickles, and compotes.

It's important that you always follow safe canning guidelines, especially with non-acidic foods. Use only disease-free fruits and vegetables, and sterilize all jars and lids before use. The National Center for Home Food Preservation website, http://nchfp.uga.edu/how/can_home.html, offers instructions for safely canning fruits and vegetables.

CURING AND STORING

Several vegetables can be cured and stored, including cabbage, garlic, onions, squash, and root vegetables such as beets, carrots, parsnips, potatoes, rutabagas, and turnips. Follow a few guidelines to make sure your stored vegetables stay fresh until you're ready to use them.

- Clean garlic and onions to remove excess soil and allow them to dry before storing. Store them at just above freezing and they will last into spring.
- Allow winter squash to mature on the vine. When a squash is ready to harvest, the vine will start to wither near the stem, and its skin should be too hard to pierce with your thumbnail. Cut it from the vine, keeping a couple of inches of stem intact, and store at room temperature. Depending on their size, squash should store for 3 to 6 months. Bruised squash should be eaten and not stored.
- Root crops can be stored for 3 or 4 months in cool, 35 to 40°F, conditions with moderate humidity. Remove the tops after harvesting. Until the ground freezes, carrots and parsnips can be left in the ground under a thick layer of mulch and harvested as needed.
- Green cabbage stores best in the cool, humid conditions favored by root crops and can last up to 6 months in storage. They can rot, however, so check on them frequently.

Dry herbs by hanging bundles upside down.

DRYING

Herbs, fruits, seeds, and popcorn can all be preserved by drying. All you need is some good air circulation, and patience.

Herbs. Leave herbs on their stems and tie them in bundles using rubber bands or string; then hang them upside down in a cool, dark place. Check on them periodically to make sure they aren't slipping through the tie. When herbs are completely dry, store them in an airtight container.

Fruit. Use a dehydrator if you have one. You can also dry fruits in the oven set at very low heat, but this method can take up to 12 hours. Fruits must be fully dried or they won't store very long.

Beans and popcorn. Dry beans and popcorn over a mesh screen until they're brittle. Then store them in airtight containers.

FREEZING

This is a quick and easy way to preserve most types of vegetables, fruits, and herbs, and it can also be the best method for retaining their flavor and nutrients. Most will last 8 to 12 weeks in the freezer. Choose airtight containers: glass and rigid plastic tend to keep out odor and moisture, but you'll find bags specifically designed for freezing that do a good job, too.

Blanche fruits and vegetables before freezing to stop the growing process. Herbs can be frozen fresh. To prevent frozen blocks of vegetables, fruits, or herb leaves, spread them out on a tray and then place them in the freezer. After they are frozen solid, add them to a container or bag. Pack them tightly to avoid freezer burn.

PLANT A ROW FOR THE HUNGRY

If you think no one wants to see another one of your zucchinis, think again. According to the Garden Writers Association (GWA), one in eight households in the United States experiences or is at risk of hunger. That's 33 million people, including 13 million children. Food banks are having a hard time keeping up with demand and many people get turned away. Fresh produce is surprisingly expensive, and your extra zucchini could help.

In 1995, the GWA Foundation created the Plant a Row for the Hungry (PAR) program, with the idea of having garden writers encourage their readers to plant an extra row of any vegetable to donate to local service organizations such as food banks and soup kitchens. Gardening groups are encouraged to form programs to coordinate these efforts. Since the program's inception, more than 2 million pounds of food per year have been donated, providing more than 60 million meals—all from individual gardens.

Local food banks will be delighted to accept your donations. Many gardeners don't think they have enough space or enough of a yield to make much of a contribution, but when you combine all those small contributions, they add up quickly. Vegetables that can be cooked or stored are especially welcome, because some crops, such as cucumbers or lettuce, can spoil before being distributed. The GWA provides tips and materials at www.gardenwriters.org to help you get started.

Cover Crops

Improving garden soil is a never-ending task. Cover crops, also called green manure, are an economical way to amend your soil before the official growing season starts. Although growing cover crops doesn't take much effort or time, you'll need to remember to sow the seed before it's too late in the season, so be sure to include cover crops in your garden plan. After you've grown cover crops and experience the soil benefits, you will become evangelical about them. Green manure crops add organic matter into the soil, much as animal manure does. But because cover crops are planted in exposed soil before you plant vegetables, they also block weeds, prevent soil erosion and compaction, and add nutrients and nitrogen.

Legumes and grasses are the cover crops of choice. Traditional legume cover crops for New York include clovers, cowpeas, soybeans, and vetches. Because legumes pull nitrogen from the air and hold onto it in their roots, they prepare the soil for heavy nitrogen users such as corn and leafy vegetables (broccoli, cabbage, cauliflower, herbs, kale, salad greens, and spinach). Legumes benefit from using Rhizobia inoculants, soil bacteria that help fix nitrogen inside legume root nodules. Inoculants should be available wherever legume seeds are sold. Of course, all plants need some nitrogen, so you can't go wrong with a legume cover.

Grasses, such as annual ryegrass and winter rye, provide organic matter and return nutrients to the soil. They grow densely and more quickly than legumes and help to suppress weeds. Mixing some grass or grain seed, such as buckwheat or oats, with your legumes offers the best qualities of both. I find grasses difficult to till into the soil, however.

COVER CROP PLANTING GUIDE

	CROP	WHEN TO SEED	WHEN TO TILL UNDER	BENEFITS
Legumes	Hairy vetch	Aug to early Sept	In spring or after flowering; may need mowing	Fixes nitrogen; winter hardy
	White clover	July to early Sept	In spring	Fixes nitrogen; winter hardy
	Cowpea	July through Aug	In spring or after flowering	Fixes nitrogen; suppresses weeds; drought resistant; not winter hardy
Grasses	Annual ryegrass	Aug through Sept	Before going to seed	Adds organic matter; grows quickly; winter hardy to zone 5
	Oats	Aug to mid-Sept	In spring	Adds organic matter; fast grower
	Winter rye	Late July through Sept	In spring; may need mowing	Adds organic matter; germinates in cool temperatures
Other	Buckwheat	May through Aug	After flowering	Adds organic matter; suppresses weeds; grows fast

Clover is a good legume cover crop.

Cover crops are easy to use: Plant. Water. Turn under or till in. You can actually plant cover crops at any time of year, except winter. You can plant them between other crops or add them to your rotation plan. Most of us, however, sow cover crops in August through September and till or turn them into the soil the following spring before they flower.

As you incorporate cover crops into your soil, don't worry about chopping them finely or digging them in deeply. Simply dislodge the roots and make sure the green tops make contact with the soil so they will begin to decompose. Try to incorporate cover crops into the soil about 3 to 4 weeks before you start planting vegetables to allow them time to decompose a bit. If the crop shows lots of top growth, you'll need to mow or cut it down before you turn it under. And winter rye, which contains germination inhibitors, needs several extra weeks to decompose before you seed new vegetables in the area.

Feeding plants midseason

Plants that continue growing and setting fruit all season expend a lot of energy doing so. Even if your soil is great, these guys are quickly using up nutrients; by the beginning of August they'll need some supplemental food. Liquid organic fertilizer is a great choice this time of year because it's faster acting than dry supplements, but you can feed your plants in a couple of ways:

Side dressing. Sprinkle 1 or 2 tablespoons of dry fertilizer near the base of the plant, keeping it at least 6 inches away from the plant's stem to avoid burning it. Any well-balanced, organic fertilizer will do. Gently work the fertilizer into the soil or let the earthworms do the work for you. Do not feed if plants are stressed or if the weather is very dry and the plants will not be getting water. And don't overdo it; too much fertilizer will build up in the soil and can harm plant roots. Some gardeners prefer to side dress throughout the season with compost or well-rotted manure, which helps to improve the soil but does not directly provide nutrients to the plants.

Foliar feeding. Leafy greens need a lot of nitrogen, which doesn't stay very long in the soil. If your greens are looking a bit chartreuse, sprinkling them with some organic fish or seaweed emulsion will work wonders. Soaking the leaves seems to quicken the results, but make sure some of the mixture reaches the roots; then continue watering regularly.

▸ Side dress plants with granular fertilizer or organic materials such as compost, rotted manure, or leaf mold.

SEPTEMBER
Chilling Out

Just when we think summer is hitting its stride, September rolls in. This is a bittersweet time in the Northeast, when the relief from the heat and the changes in nature's color palette are very welcome, but the plants seem to know their time is coming to an end. We still have plenty of crops to harvest, but the aromatic ripeness of tomatoes gives way to the hard shells of winter squash. The garden is slowing down, but that doesn't mean you get to slow down with it. Some gardeners think vegetable gardening ends with the first frost, but some vegetables are actually improved by a good chill, so you could be harvesting well into winter. And you still need to take care of a few seasonal tasks, including seed collecting and garden cleanup.

◀ Pea seeds are easy to save for next year. Let the pods dry on the vines before removing the seeds.

TO DO THIS MONTH

PLAN

- Buy garlic and shallots for fall planting
- Note reminders of what grew well, what failed, and any problems with pests
- Pull out the row covers and set up your hoop house
- Test the soil pH and add amendments for next year
- Start collecting seeds

PREPARE AND MAINTAIN

- Add manure to rhubarb bed
- Harvest remaining tomatoes to ripen indoors
- Harvest and dry popcorn

Zone 3 Zone 4 Zone 5 Zone 6 Zone 7

SOW AND PLANT

Everyone

- Sow cover crops in unused beds

Zone 7

- Direct sow: broccoli, broccoli raab, carrots, kale, lettuce, peas, and spinach

HARVESTING NOW

- Apples
- Arugula
- Beets
- Broccoli
- Broccoli raab
- Brussels sprouts
- Cabbage
- Carrots
- Cauliflower
- Celery
- Chicory
- Cranberries
- Cucumbers
- Eggplant
- Endive
- Fennel
- Grapes
- Green beans
- Green onions
- Herbs
- Kale
- Kohlrabi
- Leeks
- Lettuce
- Melons
- Onions
- Peas
- Peppers
- Potatoes
- Pumpkins
- Radishes
- Raspberries
- Rutabaga
- Shelling beans
- Spinach
- Swiss chard
- Tomatoes
- Turnips
- Winter squash

Seed Saving

When you save seeds, you bring your garden full circle—plus, you ensure that you'll be able to grow your favorite varieties next season. Each time you grow plants and save seeds from the healthiest, hardiest specimens, you are improving their adaptability in your garden. Most seed saving is as easy as letting the vegetables mature or dry on the plant and harvesting the seeds inside. To be a candidate for seed saving, the plant must be an open-pollinated variety, such as an heirloom plant. Open-pollinated varieties, often designated as OP, have stabilized over the years, so that seeds from OP vegetables will grow into plants exactly like the parent.

Hybridized varieties, on the other hand, are crosses between two different varieties and are not stabilized. Saving and planting seeds from hybrids will give you plants with some of the characteristics of the parent, but the plant won't be identical to the parent. Saving seed from hybrids can make for interesting experiments—or disappointments.

To complicate the situation a bit more, insects or the wind can cross pollinate even some OP vegetables with other varieties of the same vegetable. For example, if you're growing two types of beans on the same teepee, the bees can easily transfer pollen from one bean flower to another, and the resulting seed will be a natural hybrid. You can, however, ensure that you get seeds that will grow true to type by following one or more of these guidelines.

- Plant only one variety of each vegetable.
- Plant varieties that flower at different times, such as an early- and a late-season corn.
- Separate different varieties by the required distance. (Distance will vary by vegetable, but substantial distances are usually required to avoid cross pollination.)
- Cover the plants while they are in flower. Of course, you will need to uncover them to allow for pollination, but you could alternate, covering one variety and then the other, every other day.

HARVESTING THE SEEDS

Vegetables with seed pods that dry on the plant, such as beans, peas, lettuce, and greens, provide the easiest seeds to save. Simply allow the pods to dry completely on the plants, and then collect and clean them. Some pods tend to open and disperse seed on their own, so when the pods start to turn brown, you can pull out the entire plant and put it in a paper bag to finish drying. If the seed pods pop open, they'll fall to the bottom of the bag.

MAKING LEAF MOLD

Leaf mold is a wonderful soil amendment—and trees give you the leaves for free. Leaf mold is simply leafy compost that's created when leaves decompose into a rich, crumbly, earthy substance that can be added to soil to improve its water retention and texture. Adding leaf mold also entices all kinds of beneficial organisms to live in your soil.

Leaf mold pretty much makes itself. In fact, making it is so easy, you might find yourself eyeing bags of leaves on your neighbor's curb. But it takes about a year before you can use it.

1. Build an open pile of leaves or stuff the leaves in garbage bags. If you use a plastic bag, poke some holes in it so that air can flow through, to avoid a mucky mess inside the bag. (Shredding the leaves before you pile or bag them and turning the piles or shaking the bags every few weeks will speed up the process.)
2. Add a little water occasionally to the piled leaves, and if the bagged leaves have dried out, add a little water there.
3. Leaf mold is ready to use when it is dark and crumbly like soil and smells fresh and earthy. Add it to your garden beds and work it into the first few inches of soil, or use it as mulch.

◂ **The first step in making leaf mold is gathering a pile of fallen leaves.**

Green beans are a good choice for seed saving.

If the seeds are inside the fruit, let the fruit fully ripen well beyond the eating stage before you harvest—think of yellow cucumbers or monstrous zucchini. After the fruits have passed their peak, you can scoop out the seeds, rinse them off, and lay them out to dry on a paper plate or a sheet of paper.

STORING SEEDS

After harvesting seeds, you'll need to prepare them for storage so they will last and remain viable for next year. Seeds need to be dry to the point of being brittle before you store them away or they can rot. Preparation of seeds involves a few steps:

1. Clean away as much of the chaff as possible.
2. Put dry seeds in paper envelopes, and place the envelopes inside an airtight container, such as a canning jar.
3. Store the container in a cool, dark spot.

Next year's crop depends entirely on this year's seed, so always choose the best quality plants—those with great flavor and in perfect health—for the best seeds. And try to use the seeds within one to two years.

Getting green tomatoes to ripen

Why, oh why, do tomato plants keep producing tomatoes when they clearly have no time to ripen? Sure, you can cover them for a few weeks and protect them from frost, but tomatoes need heat and sun to ripen and develop their mouth-watering sugars and flavor. I refuse to surrender to green tomatoes, so I've developed a few tips for getting the stubborn fruits to ripen off the vine.

First, when night temperatures stay in the low 70s°F, remove all the flowers from your tomato plants. You may find this difficult, but remember that they aren't going to mature into fruits, and they're sapping energy from the existing fruits. Later, when frost is imminent, pick the tomatoes that show promise of ripening and bring them indoors. Mature fruits on the precipice of ripeness will show a blush of color on their blossom (bottom) end and will have just started to soften. At this point, you have several ripening options.

- Lift the whole plant out of the ground, roots and all. Shake off the excess soil, but leave the roots intact, and hang the whole plant upside down in a protected location with indirect sunlight. The fruits will continue ripening on the vine. This is my favorite method, because it allows the tomatoes to develop vine-ripened flavor.
- Put the tomatoes on a sunny windowsill. This isn't the most reliable method, but it's easy.
- Place the tomatoes in a paper bag with an apple, which gives off ethylene gas and encourages ripening.
- Wrap each tomato in newspaper and store them in a dark, dry spot. Check them periodically to make sure they are not rotting, and they should be ready to eat within 3 to 4 weeks.

OCTOBER

Getting Ready for Bed

The scent and crunch of fallen leaves under foot, and no bugs, make midautumn a wonderful time to be out in the garden. We can take pleasure in the crisp, breezy days and low, glowing light. October is a great time to tackle all those chores you put off, such as testing your soil and turning the compost pile. If you've been thinking about putting up a hoop house, get moving. Get the garden ready for bed. Toss the old plants and debris into the compost and harvest the perishables. Keep your camera or notebook handy and make some end-of-season notes about what you want to do first thing next spring. The main event is over, but there's still a lot to do.

◀ Learning to cure and store winter squash correctly will keep you in veggies all winter long.

TO DO THIS MONTH

PLAN

- Make notes in your garden journal (weather, late season crops, harvest dates)

PREPARE AND MAINTAIN

- Get frost covers ready
- Last chance to put up a hoop house
- Clean garden of all debris
- Rake and pile leaves for leaf mold
- Harvest pumpkins and winter squash before a frost hits
- Cure winter squash for storage

SOW AND PLANT

Everyone

- Finish planting garlic and shallots
- Pot up annual herbs to bring indoors

Zone 7

- Direct sow: beets, broccoli, carrots, cover crops, kale, lettuce, and spinach

HARVESTING NOW

- Apples
- Arugula
- Beets
- Broccoli
- Brussels sprouts
- Cabbage
- Carrots
- Cauliflower
- Celery
- Cranberries
- Cucumbers
- Eggplants
- Fennel
- Grapes
- Herbs
- Kale
- Kohlrabi
- Leeks
- Lettuce
- Melons
- Onions
- Parsnips
- Peas
- Peppers
- Potatoes
- Pumpkins
- Rutabagas
- Shelling beans
- Swiss chard
- Winter squash

Zone 3 Zone 4 Zone 5 Zone 6 Zone 7

Curing and Storing Late-Harvest Vegetables

By now, even the longest season winter squash should be harvested and ready for storage. But before you tuck them away, they'll need a little prep work. Winter squash, along with potatoes and sweet potatoes, are still alive and growing even after being harvested, and you need to send them into dormancy. They must be cured to prevent them from rotting and to make sure every last treasure eventually makes it to the table. Curing allows the vegetables to develop a hard, protective outer skin or shell.

Curing isn't difficult to do, but it is crucial. In a warm, airy, and dry spot away from direct sunlight, spread out the vegetables in a single layer. If you have some old window screens, they make the perfect venue. Lay the screens over a couple of saw horses or bricks and let the vegetables sit for a couple of weeks. To save space, stack several screens on top of one another like shelves, separated with bricks or cinder blocks.

Cool temperatures prolong the life of stored vegetables. It's a bit of a balancing act, though, because if it gets too cold, the vegetables will die and rot. Humidity also plays a part in successful storage; some humidity is needed to keep the vegetables from drying up and shriveling, but too much will cause them to mold or rot.

STORING WINTER SQUASH

Keep in mind that frost and prolonged cold temperature can damage squash in the garden. Harvest or cover the plants if a frost is forecast or if the temperature is going to dip below 40°F. Squash won't be worth storing if it is exposed to these conditions, so you might as well harvest and eat what you can, and prepare, package, and freeze the rest.

Immature squash will not last long and should be eaten within a few weeks, so before you even think about storing squash, you need to make sure they are fully mature. Winter squash is fully mature when you can no longer puncture the skin with your thumbnail. When fully mature, winter squash go through subtle changes in appearance. Pumpkins should be completely orange, with no green showing. When ripe, the surface of the squash, even a pumpkin, will begin to dull, losing its glossy sheen. The vines will also start to die off.

After you've identified and harvested the mature fruits, inspect them all for signs of damage. Any ding in the skin could be an invitation for rot. You'll be curing and storing only healthy, uninjured fruits. Wash the squash with a solution of 1 part household bleach to 10 parts water to remove any clinging soil and surface pathogens. Then dry them well and set them aside to cure.

Knowing a few tips will help in storing squash.

- Harvest when the plants are dry, and keep the squash dry during the curing process.
- Store only uninjured, healthy squash.
- Leave 3 to 4 inches of stem on pumpkins and about 1 inch of stem on other squash.
- Don't use stems as handles, because they can easily break and cause cracks in the fruit.
- Handle squash carefully to avoid scratching the skin surface.
- Leave plenty of space so the squash are not touching one another.

WINTER SQUASH STORAGE INFORMATION

TYPE OF SQUASH	TEMPERATURE	HUMIDITY LEVEL	STORAGE TIME
Acorn	50°F	50 to 75 percent	5 to 8 weeks
Butternut or buttercup	50°F	50 to 75 percent	2 to 3 months
Hubbard	50 to 55°F	70 to 75 percent	5 to 6 months
Pumpkin	50 to 55°F	50 to 75 percent	2 to 3 months

- Store squash away from apples, which emit ethylene gas.
- Check squash often and remove any that show signs of rotting.

STORING WHITE POTATOES

By September or October, the tops of your potatoes will be dying back—it's time to dig them up and bring them in. Try to do this after a series of dry days and don't water before digging to make the harvest easier. The newly dug potatoes will be a bit damp, but they'll dry within a couple of hours and you'll be able to brush off excess soil. Don't wash or hose them off, because potatoes take a long time to dry.

Once the potatoes are clean and dry, move them to a dark holding area where they can remain for 1 to 2 weeks to cure, which seals the potato skins and heals any nicks. They cure best at 55 to 60°F and high humidity (85 to 95 percent). Store the cured potatoes in a cool (40 to 45°F), dark place with good ventilation and high humidity to prevent the spuds from sprouting. Check the spuds often for signs of rot and remove any that look suspect. They should last for up to 8 months.

If the potatoes are exposed to light, their skins will turn green. Light causes the development of solanine, which is slightly toxic. Eating a large helping of green potatoes could make you sick, but most of the solanine is just below the surface of the skin, so you can cut away the green bits before cooking and eat what remains.

STORING SWEET POTATOES

Sweet potatoes are cured to improve their texture and heighten their sweetness. Fresh sweet potatoes tend to be dry and starchy; curing them after harvesting turns them into the moist comfort food that is so good in pies. These warm-season vegetables should be harvested before frost threatens; frost will kill the vines, and cold soil or a hard frost can damage the tuberous roots, making them bad candidates for storage. As with white potatoes, try to harvest sweet potatoes when the plants and tubers are dry.

After harvest, sort through your haul and remove any damaged potatoes. After they are clean and dry, cure the remaining crop for 2 to 3 weeks at 85 to 90°F. Like white potatoes, sweet potatoes need high humidity during curing. You can increase the humidity by curing them in plastic bags, but first be sure to poke lots of holes in the bags to prevent condensation from forming, which would encourage bacteria growth. After curing sweet potatoes, store them in a dark, dry spot at 55 to 60°F.

Make your own sweet potato slips

While you are storing away your cured sweet potatoes, put a few aside to make new shoots, called slips, for planting next season. If you aren't saving your own homegrown sweet potatoes, order certified disease-free seed potatoes. Much like regular potatoes, sweet potatoes are grown from pieces of the tuber that have been coaxed into rooting and sprouting new leaves. Start rooting your slips about 4 to 6 weeks before your last frost date. Plan on transplanting them outdoors in mid-May to mid-June, when the soil feels warm to the touch.

STEPS:

1. Slice the sweet potato in half lengthwise.
2. Lay out the pieces on dampened potting mix and cover with a couple of inches of soil.
3. Loosely cover the container with plastic to keep the soil moist and warm.
4. Roots will start forming first, and within 5 to 6 weeks, you should see leaves sprouting.
5. When the plants are 4 to 6 inches tall, they're ready to transplant.

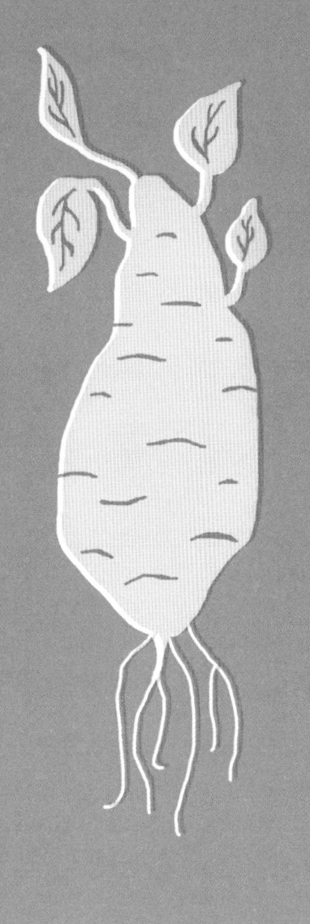

NOVEMBER
Celebrating the Harvest

The leaves are down for the count and the compost pile is too solid to turn. Time to sharpen the tools and hang them up for the season. Make sure the garden is tidy and ready to hibernate under snow cover. Clean up debris, and pick up fallen fruits and seed pods on the ground, which can invite trouble such as gnawing animals and overwintering diseases. The garden has been quiet for a month or so, but if you're lucky, this last month of fall will offer a few warm days, and the hardy greens you've left in the ground—just in case—can be cut and enjoyed for dinner. It's like cheating Mother Nature and getting a bonus month.

◀ Crops like kale happily survive a frost, providing you with a late-season harvest.

TO DO THIS MONTH

PLAN

- Review your notes while your garden experiences are still fresh in your mind

PREPARE AND MAINTAIN

- Finish harvesting and curing
- Amend the soil
- Mulch around perennial vegetables
- Clean, sharpen, and oil tools
- Check on the hoop house

HARVESTING NOW

- Beets
- Broccoli
- Brussels sprouts
- Carrots
- Cauliflower
- Celeriac
- Fennel
- Herbs
- Kale
- Leeks
- Pumpkins
- Rutabagas
- Shelling beans
- Swiss chard
- Turnips
- Winter squash

Zone 3 Zone 4 Zone 5 Zone 6 Zone 7

Some Vegetables Can Take the Cold

Tender, heat-loving vegetables such as tomatoes and peppers will turn to mush with the first kiss of frost, but many others will continue on undaunted. Generally, there are a few light frosts before a hard, killing frost brings the season to a close. A light frost hovers around the freezing point, at about 30 to 32°F. When the temperature dips down below 28°F, what doesn't die goes dormant.

If you've set up a hoop house or cold frame, or your row covers are handy, you can protect your garden from the first few frosts. While inside a cold frame or hoop house, many plants that tolerate hard frosts will still be harvestable through the winter, provided you can get to them through the snow and the ground is not frozen solid.

Along with the dipping temperatures, a couple of additional factors will affect your vegetables' survival. The ground holds heat longer than the air, so vegetables growing

FROST-HARDY VEGETABLES

TOLERATE A LIGHT FROST

- Arugula
- Asian greens
- Beets
- Carrots
- Cauliflower
- Celery
- Endive
- Lettuce
- Mache
- Parsnips
- Peas
- Radicchio
- Swiss chard

TOLERATE A HARD FROST

- Broccoli
- Brussels sprouts
- Cabbage
- Collards
- Kale
- Kohlrabi
- Leeks
- Mustard greens
- Onions
- Parsley
- Radishes
- Spinach
- Turnips

near soil level and root crops safely tucked underground will be less likely to be harmed by the cold. Humidity also protects plants, so watering your plants the night before an expected frost can help insulate them from damage. On the other hand, wind intensifies frost damage, so for an exposed garden, cover is the best option.

Mulching Around Perennials

Perennial vegetables, such as asparagus, artichokes, and rhubarb, take a year or two to become established, but they'll give you many, many years of payback. Most seem to take care of themselves, but a little help is always welcome. These vegetables are a hardy bunch, but they enjoy a blanket of protective mulch in the winter.

ASPARAGUS

You can cut back asparagus fronds in late fall or early spring. I opt for late fall, when they have fallen over and have been killed by frost, because I don't want pests making a home for the winter in there. To avoid exposing the crowns to freezing and thawing, after the ground has frozen, spread 4 to 6 inches of chopped leaves or straw over the entire bed. In early spring, when the ground has started to thaw, you can remove the mulch. Or you can remove the mulch from half the bed, and because the exposed soil will heat up first, half of the asparagus will come up earlier than the rest. Once they pop out of the ground, remove the remaining mulch to encourage the rest of the asparagus to grow and extend the harvest. Don't leave the mulch on top of emerging plants or they won't grow straight.

ARTICHOKES

Gardeners in zone 7 will be able to overwinter artichoke plants in the garden without a lot of protection. Artichokes can take a slight frost, but prolonged freezing will kill their roots. If you want to try overwintering plants in colder areas, start by cutting them back to about 6 inches, and then mulch with 6 inches of packed leaves or straw. I like to place an open box on top of the plant and pack the leaves inside. I put rocks or bricks on top of the flaps on the ground to hold the box in place. Once it is full of leaves, I tuck the box top closed. I leave the box like this until a few weeks before my last frost date, and then remove it and cross my fingers.

RHUBARB

Rhubarb actually seems to like the cold weather, and I don't worry about keeping it mulched so much as keeping it fed. Fall is the best time to divide an established clump—something I'd recommend you do faithfully every couple of years. Once rhubarb becomes overgrown, it is almost impossible to cut through. But if you catch it early enough, you stand a chance of getting your shovel blade through the center of the plant. Slice around and remove the center of the plant, leaving the periphery in place. If the shovel won't go in, try using a garden knife. Then fill the center area with compost or other organic material and let it work its wonders through the winter.

Rhubarb is a perennial edible that is not bothered by cold weather.

Root Cellars and Storage

If you can provide a cool, dark, dry location, you can store many of your harvested fruits and vegetables during the winter months. Root cellars are a step back in time, because they were the only storage option before electricity and refrigeration were available. Generally built into the ground or a hillside, root cellars use the natural insulation of the surrounding earth to maintain a low, but not freezing, temperature and relatively high humidity.

Many vegetables and tree fruits will store well in a root cellar.

- Root vegetables: beets, carrots, garlic, onions, parsnips, potatoes, rutabagas, and turnips
- Cabbage
- Pumpkins and winter squash
- Tree fruits (store apples separately, because the ethylene gas they release will cause other vegetables to spoil)

If you want to create a root cellar, keep in mind three important considerations: temperature, humidity, and air flow. If you decide to build a large cellar or make storage space inside your home, get a hygrometer to measure humidity and a thermometer that records both the highest and lowest temperatures. If you are storing vegetables with different needs, you can compensate by keeping those that need the highest humidity levels near the floor, which tends to be the dampest area.

If an in-ground root cellar seems like a luxury, you can try a couple of relatively simple, quick, and inexpensive ideas for storing your harvest. The only question is how big you want to go.

THE BURIED BUCKET OR CAN CELLAR

To create a DIY in-ground root cellar, start with a 5-gallon plastic bucket. This is a small cellar, to be sure, but it works well for root vegetables—and you can always bury multiple buckets. Dig a hole large enough to accommodate the bucket except for the top 3 or 4 inches, which should remain above ground level. Fill the bucket with vegetables and close the lid. Add some clamps to keep it closed and a small tarp on top to keep water from seeping in. Then mark the area with a post and flag, so you can find the bucket in the snow and won't trip over it later.

You can make a larger version using a clean garbage can, although the larger hole is considerably more difficult to dig. I prefer the galvanized steel cans because they don't

▶ **Make sure your bucket, or any root cellar, is well sealed to keep mice and other hungry animals out.**

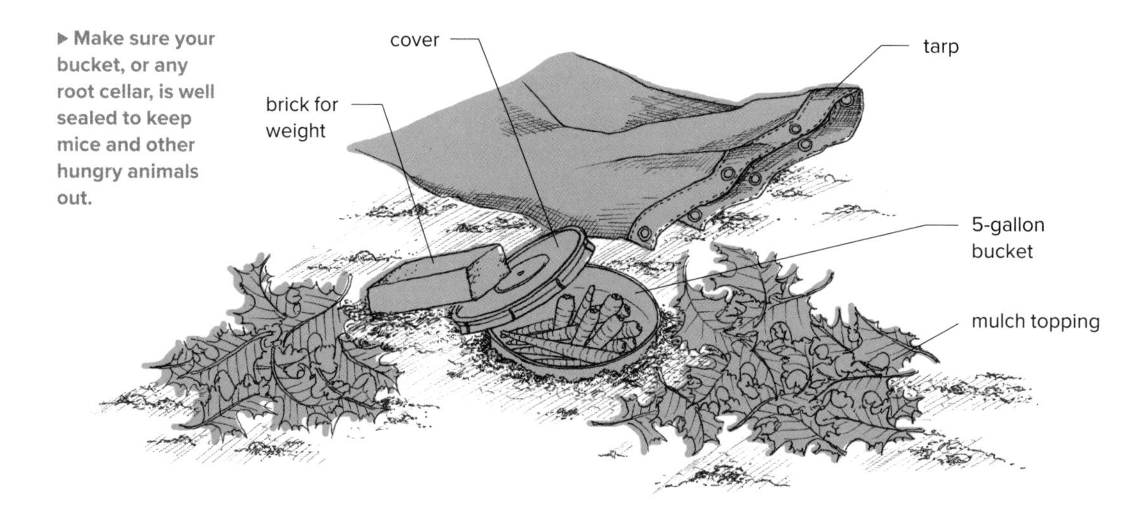

crack. Drill five or six ¼-inch holes around the bottom of the can to allow a little humidity in and prevent the vegetables from drying out. Few insects are active in the winter, but just in case, you can line the bottom with window screening to keep them out. Leave 3 or 4 inches of the can above soil level, and use either clamps or a bungee cord to keep it closed and a tarp to avoid water seepage. Store the vegetables in net sacks to make retrieving them from the bottom of the can a little easier.

In areas with harsh winters, you'll find it easier to open a bucket or can cellar if you cover the top with 8 to 12 inches of leaves or straw before you add the tarp—this helps prevent the cover from freezing onto the can.

THE BASEMENT CELLAR

If you have an unheated basement or cellar in your home, you might want to move the baseball cards and old clothes to another room and make space for an easy and accessible root cellar.

Choose a space next to an outdoor-facing wall, preferably with a northern exposure. You'll need a vent to the outside, and the easiest way to do this is to choose a wall with a window. All of the pipes and fittings you'll need are available in the plumbing aisle at the hardware store: a piece of plywood, 1½-inch PVC pipes and fittings, a bit of screening, and some foam insulation.

1. Remove the glass in the window and replace it with a piece of plywood cut to fit.
2. Drill two 1¾-inch holes in the plywood to install vents: one to allow cold air in and one to vent the warm air out.
3. Use 1½-inch PVC pipe for the vents. The pipes need to extend only a few inches outside and inside the plywood window. One short pipe will be situated at window height, venting out the warm air that rises. On the second pipe, attach an elbow joint and a longer piece that extends down to about 12 inches above the floor.

WHAT TO STORE IN A ROOT CELLAR

	TEMPERATURE	HUMIDITY	SHELF LIFE
Apples	33 to 40°F	80 to 85%	3 to 6 months
Beets, carrots, parsnips, turnips	33 to 40°F	90 to 95%	3 to 6 months
Broccoli, cauliflower	33 to 40°F	85 to 90%	1 month
Brussels sprouts, cabbage, celery, kohlrabi	33 to 40°F	90 to 95%	1 to 4 months
Leeks	33 to 40°F	80 to 85%	3 to 4 months
Onions, garlic	40 to 50°F	55 to 70 %	6 to 8 months
Potatoes	33 to 40°F	80 to 90%	6 to 8 months
Pumpkins, winter squash	45 to 50°F	55 to 60%	3 to 4 months

This pipe will funnel cold air into the room. (Remember that warm air rises, and cool air sinks.) The airflow from the two pipes will create a siphoning effect.

4. To make sure only air gets in, cover the outside pipe ends with screening and secure that in place with couplings.
5. On the inside ends, either install blast valves or use pipe caps to close the vents whenever the temperature drops below freezing.
6. Finish off by sealing the openings in the plywood around the pipes and edges with foam insulation.

AN IN-GROUND CELLAR

Earth is a wonderful insulator. It can take some work to dig an in-ground cellar, but you'll have to do it only once. Root cellars are basically large holes dug into the ground or into the side of a hill. The sides are lined with material to keep them from caving in—rocks, wood, cinder blocks, or similar materials. A concrete floor with footings below the frost line will keep the ground from heaving. Once the cellar is lined, you'll need to top it with a door, which makes for the easiest access, or you can use a simple board covered with a tarp.

Some type of venting is needed to moderate the temperature. Two lengths of 3-inch-wide PVC pipes will do the job: One venting pipe should start near floor level to bring the cold air in. A venting pipe situated near the top, where heat rises, will vent the warmer air out. Screen the vents to prevent unwanted visitors. These are just the basics, but if you are building down, you'll need some type of ladder or stairs to get in and out, and you'll also need shelves.

Vermicomposting: composting with worms

In cold climates, winter often brings a halt to outdoor composting, because it's impossible for a frozen pile to decompose. But we still have lots of kitchen scraps, so we might as well find a use for them. How about indoor composting with worms?

Now, now, don't wrinkle your nose. These are very cultivated worms. Worm composting, also known as vermiculture or vermicomposting, is a speedy way to create compost in small bins. It involves the same basic setup required for regular composting, but on a smaller scale, and the worms are doing all the work so you don't have to turn the pile. You feed the worms; they eat, digest, and leave their lovely, nutrient rich droppings behind.

Two types of worms are best for vermicomposting: red wigglers and red earthworms. You can order these online or purchase them from a garden center. Don't use worms you've dug from your yard. Earthworms and night crawlers from your garden do not adjust well to living in a bin; they need to move through soil and feed on organic matter that is already partially decomposed. Red wigglers and red earthworms, on the other hand, thrive in compost and quickly break down organic matter. You can start by ordering a batch of each.

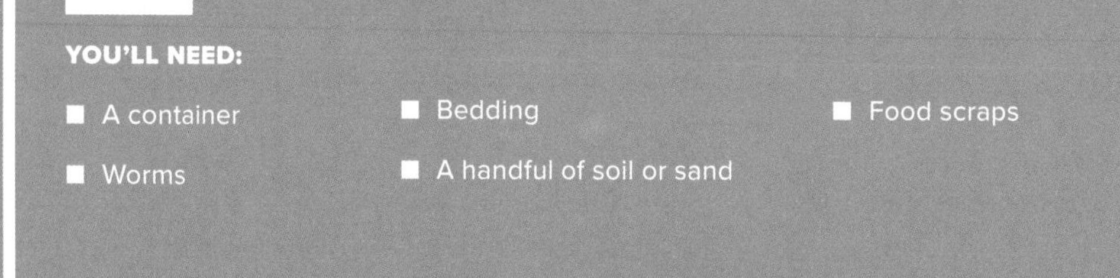

YOU'LL NEED:

- A container
- Worms
- Bedding
- A handful of soil or sand
- Food scraps

CHOOSING A BIN

Almost any container that allows for airflow and drainage will do: wood and plastic are two good materials to consider. Metal conducts heat and can get too hot or cold for the worms. Styrofoam might seem like a good idea, but it can be toxic to the worms. Many commercial worm bins are available in garden centers and online, and these are often the fastest and easiest choice.

SETTING UP A BIN

A basic, single-bin system is easy to set up. If you're using a plastic storage bin, make sure it has a cover. Drill a few small holes on the sides an inch or so below the lid for the required air flow. To collect the compost, you'll need to empty out the worms and other materials, so figure this into your setup. To make collection easier, build or purchase a stacked worm bin system that includes trays, with openings on the bottom, stacked one on top of another. As you add layers of food and bedding, the worms feed and work their way to the upper trays. Once the creatures have moved up a level, you can remove a lower tray and remove the compost. A variation on stacked bins is a horizontal layout, which works on the same principle: the worms move down the line as they finish eating the food in the preceding layer.

GETTING IT GOING

You'll need to add some bedding for the worms. Use materials high in carbon, such as shredded newspaper or cardboard, chopped up hay or dry leaves, or coir (made from coconuts). Moisten the bedding so that it's damp, but not dripping, and then toss in a handful of soil or sand. Worms have no teeth and soil provides grit that helps them digest their food. Finally, add the worms.

Every day, feed the worms vegetable scraps at a rate of about half of the worms' combined weight. So, for example, if you have 1 pound of worms in the bin, you'd feed them a half pound of scraps each day. You can adjust the amount of food scraps you provide, but don't add too much new food until they have pretty much finished gobbling what is there.

If your bin starts to smell, it's probably too wet and/or the airflow is insufficient. Cut down on the food scraps, add some more dry bedding material, and stir things up a bit to get some fresh air into the mix. The odor should soon disappear. Although worms will eat meat scraps, feeding them only vegetables and fruits will decrease potential problems with odors, flies, and rodents.

HOW TO USE WORM COMPOST

What does worm compost do for your soil? In addition to adding nutrients, it contains large populations of beneficial microbes that help make soil nutrients more accessible to plants. Worm compost is a very concentrated soil conditioner. You can use it as a side dressing in the garden or mix it into your potting mix at a rate of 1 part worm compost to 4 parts mix. Don't use it alone for starting seed, because it's not sterile and can cause problems with damping-off disease.

DECEMBER
Rest, Reflect, Rejoice

The harvest season is a memory now. Thank goodness you took plenty of notes along the way, so you can feel free to shift focus for a while. You're probably busy with holiday planning, cooking, and wrapping gifts from your garden. Summer seems an eternity away, but you'll be getting back in gear next month, because seed catalogs are arriving and you're making room for seed-starting supplies. Feel free to tuck garden planning in the back of your mind, and take advantage of this brief respite to enjoy the other passions of your life. Let your mental garden clutter sort itself out so you can breeze through the holiday season and return with visions of sugar beets dancing in your head.

◀ A garden beside Lake Ontario in upstate New York rests until next season.

TO DO THIS MONTH

PLAN

- Order catalogs
- Note weather conditions in your garden journal

PREPARE AND MAINTAIN

- Check on vegetables in storage
- Check on vegetables in hoop house

HARVESTING NOW

- Arugula
- Beets
- Brussels sprouts
- Carrots
- Evergreen herbs
- Kale
- Leeks
- Parsnips

Zone 3 Zone 4 Zone 5 Zone 6 Zone 7

Enjoying the Garden in Black and White

Winter brings a wonderful gift to cold-climate gardeners: we get to see the bones of our gardens, without the distraction of leafy plants. We usually think of garden bones when designing ornamental gardens, but if you'll be living with and in your vegetable garden for seasons to come, shouldn't it be pretty, too? When snow blankets the ground, only the lines of the garden are visible. As you start your garden dreams for next year, animate that black-and-white garden and make plans to turn it into a destination that will satisfy your eyes as much as your appetite.

This is one garden chore you can do standing inside at the window, with a hot drink, listening to inspiring music. What do you see? Is the fence merely a means to keep Peter Rabbit out, or could it be used to trellis lemon cucumbers or an espaliered pear? What about a window box of edible flowers on the gate? Do you like rearranging the garden layout every year, or would a geometric pattern of raised beds bring order and harmony to the garden? Why not make your vertical trellises an accent, even before the bean vines cover them with flowers?

Heirloom Fruits and Vegetables

It seems the new seed catalogs keep coming earlier every year. In December, I am usually too busy with other things to think about placing a seed order—and it's just as well, because I would order way too much if I did it on impulse. I look forward to all those old favorites in my garden, especially green beans and hot peppers, but I always make room to try something new.

I have become an heirloom vegetable zealot. It all started with the quest for the perfect tomato. I had no idea where it would take me. If you have grown or even tasted heirloom tomatoes, you know that the flavor and tenderness are like nothing you can buy in a grocery store. Take that revelation and expand it into a world of flavors and choices, and you have literally thousands of heirloom fruits and vegetables to try growing.

Most heirloom fruits and vegetables are not mass produced, and many are too delicate to make the trip from farm to market. These are varieties that gardeners have saved and passed along for many years—and you know that gardeners wouldn't save and grow plants that didn't taste terrific. So if you long for the bite of a melon that melts into sugar, squash that can perfume the whole house while it's baking, or a tender-fleshed eggplant to caramelize on the grill, focus your attention on the catalog varieties marked as heirloom or open pollinated (OP). Better still, look for heirlooms noted as good varieties for the Northeast.

Because so many seed companies got their starts in this part of the country, you can imagine how many varieties are suited to our growing conditions. I'm thinking about 'King of the North' sweet peppers, with a short growing season that means they'll fully ripen to red in time to harvest and enjoy. Or the pie-sized 'Connecticut Field' pumpkin; the wheel-shaped 'Long Island Cheese' pumpkin, a stringless winter squash that has been favored for generations; and 'Schoon's Hardshell' melon, a native New Yorker that delivers huge, late-season melons, well worth the wait. Fruit lovers can try the Massachusetts cider and pie 'Baldwin' apple, the 'Magness' dessert pear from Maryland, and the classic highbush 'Jersey' blueberry. We don't need to make do with the quirky growing season; we can embrace it and grow great varieties that thrive here.

Gifts from the garden

Gardeners love to share from their gardens during the growing season, and there's no reason why we can't keep giving during the winter. If you've been drying herbs and flowers or making jams, they'd make cheerful gifts during the bleak days of winter. I'll often bring a plant I've grown from cuttings or a bottle of herb-infused oil or vinegar as a hostess gift for holiday parties. I'm sure you have a friend or two who admired your tomatoes or gourds; they'd probably be delighted to be able to grow their own next year. If you've saved the seeds, you can share them as gifts. You can also create herb mixes and even herbed spirits from your garden herbs.

MAKE HERBAL VODKA

Now here's a hostess gift that keeps on giving. Don't worry; you don't have to make the vodka from your own potatoes. Just be sure to buy a good quality, unflavored variety. To store this infusion, use any glass container with a lid. I generally use several mason jars, but you can use a large container if you plan to make a lot of one flavor. Rub your herb sprigs gently to release their oils, and then toss them into the jar. Pour in the vodka and close the lid tightly. Give the jar a good shake daily; the herbs will infuse the vodka in 2 or 3 days. If you want to add some berries or seeds, give it a week.

After a few days, test a sip. When you can taste the herbs to your satisfaction, strain out the greenery and transfer the vodka to the gift bottle. Label it and tie some fresh sprigs around the bottle neck.

Infused vodka can be kept refrigerated for up to a month, or it can be kept in the freezer for up to 2 months. Favorite herbs for vodka include basil, lavender, lemon balm, mint, tarragon, and thyme. Consider adding chili peppers, citrus peel, cardamom seeds (really nice with mint), or some festively colored cranberries. Cheers!

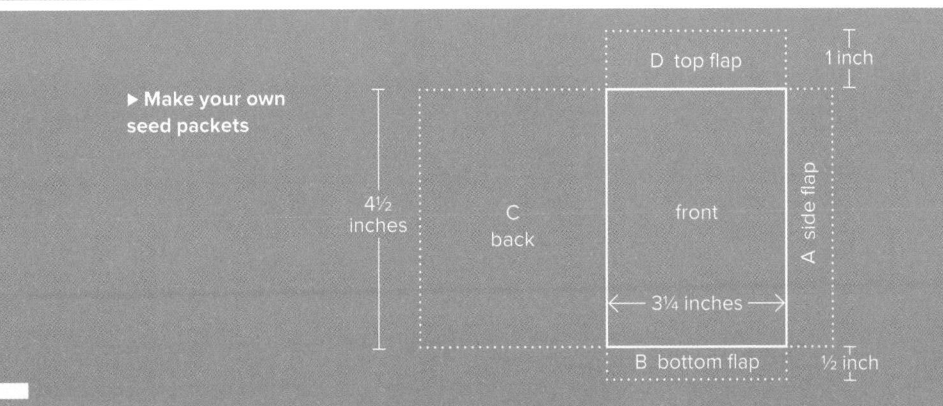

MAKE YOUR OWN SEED PACKETS

Saving your own seeds is such a personal, sensory thing to do, and it seems a shame to store them in plain, old envelopes, especially if you are going to be sharing them. Homemade seed packets are easy to assemble and a lot of fun to decorate and share. You can personalize packets with photos of the plant in your garden and some growing instructions. Use the envelopes as holiday cards or turn them into gift tags or hanging ornaments by punching a hole in one corner and threading a ribbon through. Tuck several in a pot or a small watering can. For friends who love to cook, add packets to a serving bowl, along with some recipes. Sharing seed packets with photos and notes is almost like sharing your garden journal. If you start a seed exchange tradition with friends, you will have a lifelong record of your gardens.

Using a template and decorative paper, you can make your own seed packets. I keep a template handy on my computer, so I can type in the plant name and some growing tips before printing.

STEPS:

1. Trace or sketch a box like the one shown with dimensions listed.
2. Cut along the dotted lines.
3. Fold along the solid lines: Fold side flap A, and then fold bottom flap B.
4. Apply glue to the outside of flaps A and B, and then fold the back, C, on top of them. Press firmly and allow the glue to dry.
5. Pour in your seeds, and fold down and tuck in or glue top flap D.

MAKE YOUR OWN GOURMET HERB MIXES

Did you harvest and dry all those wonderful herbs before the first frost? Why not whip up your own Herbs de New York blend? Have you ever met a vegetable gardener who didn't like to cook? You can mix up your favorite blend and put it in a canister or jar to give to friends. Don't forget to attach a recipe card. Here's my house blend, to get you thinking.

SUMMER MEMORY BLEND

- 3 Tbsp. dried oregano
- 3 Tbsp. dried thyme
- 1 Tbsp. dried marjoram
- 1 Tbsp. dried basil
- 2 tsp. lavender flowers
- 1 tsp. dried rosemary
- 1 bay leaf

Edibles A to Z

Apples, beets, carrots—nothing compares to the ABCs of your own garden.

PLANTING AND HARVESTING CHART

These charts show the planting and harvesting periods for a variety of edibles. You'll find more information about starting seeds indoors, which vegetables to direct sow and succession plant, and planning for fall, in the following profiles of specific vegetables. When to start planting is a balancing act between average frost dates, yearly weather fluctuations, and our own impatience. While it is tempting to start seeds early, you won't gain any advantage if the weather does not cooperate. The dates on these charts are a good guideline, but you will need to be prepared to make seasonal adjustments.

ZONE 3 AND 4

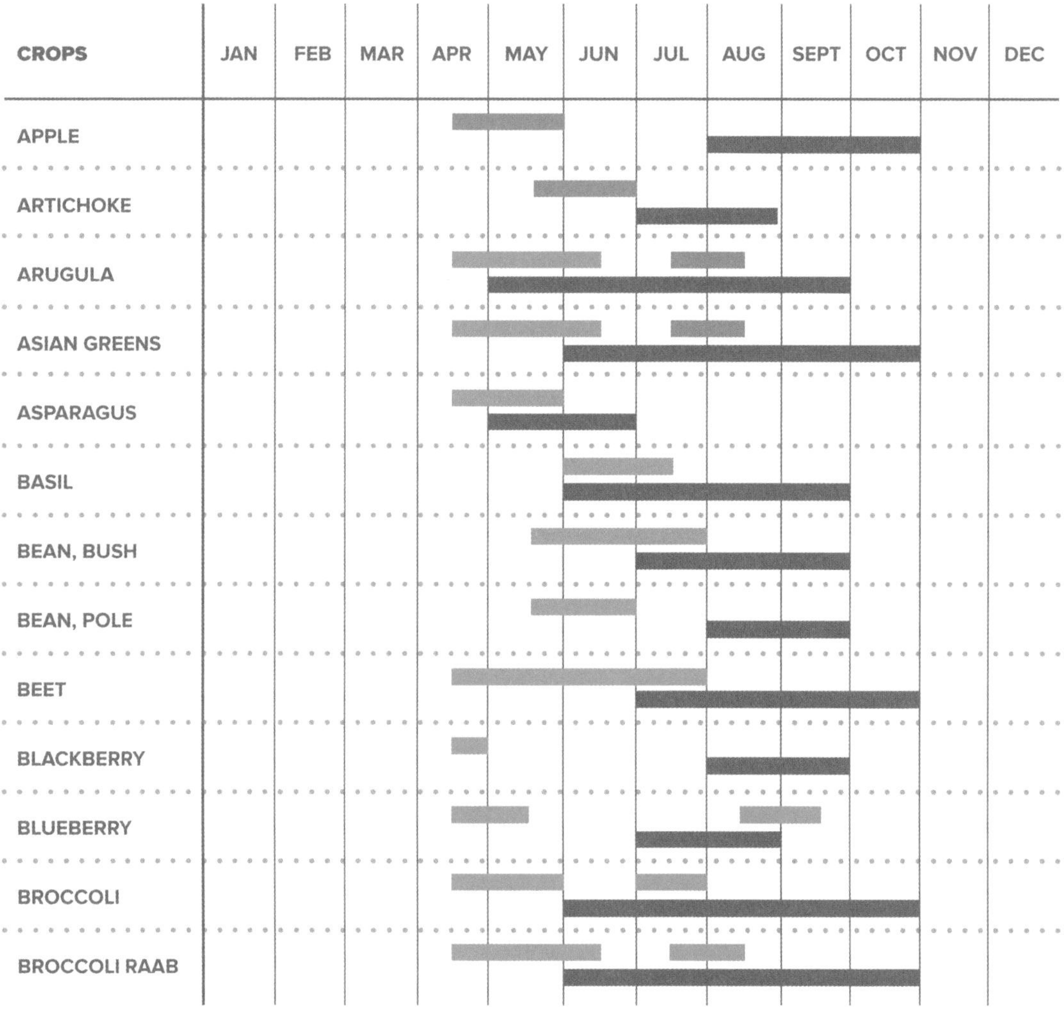

CROPS	JAN	FEB	MAR	APR	MAY	JUN	JUL	AUG	SEPT	OCT	NOV	DEC
APPLE												
ARTICHOKE												
ARUGULA												
ASIAN GREENS												
ASPARAGUS												
BASIL												
BEAN, BUSH												
BEAN, POLE												
BEET												
BLACKBERRY												
BLUEBERRY												
BROCCOLI												
BROCCOLI RAAB												

CROPS	JAN	FEB	MAR	APR	MAY	JUN	JUL	AUG	SEPT	OCT	NOV	DEC
BRUSSELS SPROUTS				Planting	Planting				Harvesting	Harvesting	Harvesting	
CABBAGE				Planting	Planting		Harvesting	Harvesting	Harvesting	Harvesting		
CARROT				Planting	Planting	Planting, Harvesting	Harvesting	Planting, Harvesting	Harvesting	Harvesting	Harvesting	
CAULIFLOWER				Planting	Planting			Harvesting	Harvesting			
CELERY/CELERIAC					Planting			Harvesting	Harvesting			
CHERRY				Planting	Planting		Harvesting					
CHICORY				Planting	Planting			Harvesting	Harvesting			
CILANTRO				Planting	Planting, Harvesting	Harvesting		Planting	Harvesting	Harvesting		
CORN					Planting	Planting	Harvesting	Harvesting	Harvesting			
CUCUMBER					Planting	Planting	Harvesting	Harvesting	Harvesting			
CURRANT					Planting	Harvesting	Harvesting					
EGGPLANT						Planting	Harvesting	Harvesting				
ENDIVE				Planting	Planting			Harvesting	Harvesting			
FAVA BEAN				Planting	Planting, Harvesting	Harvesting						
FENNEL					Planting	Planting	Planting	Harvesting	Harvesting			
GARLIC							Harvesting	Harvesting		Planting	Planting	
GOOSEBERRY					Planting	Harvesting	Harvesting					

CONTINUED ▸

PLANTING AND HARVESTING CHART

ZONE 3 AND 4, CONTINUED

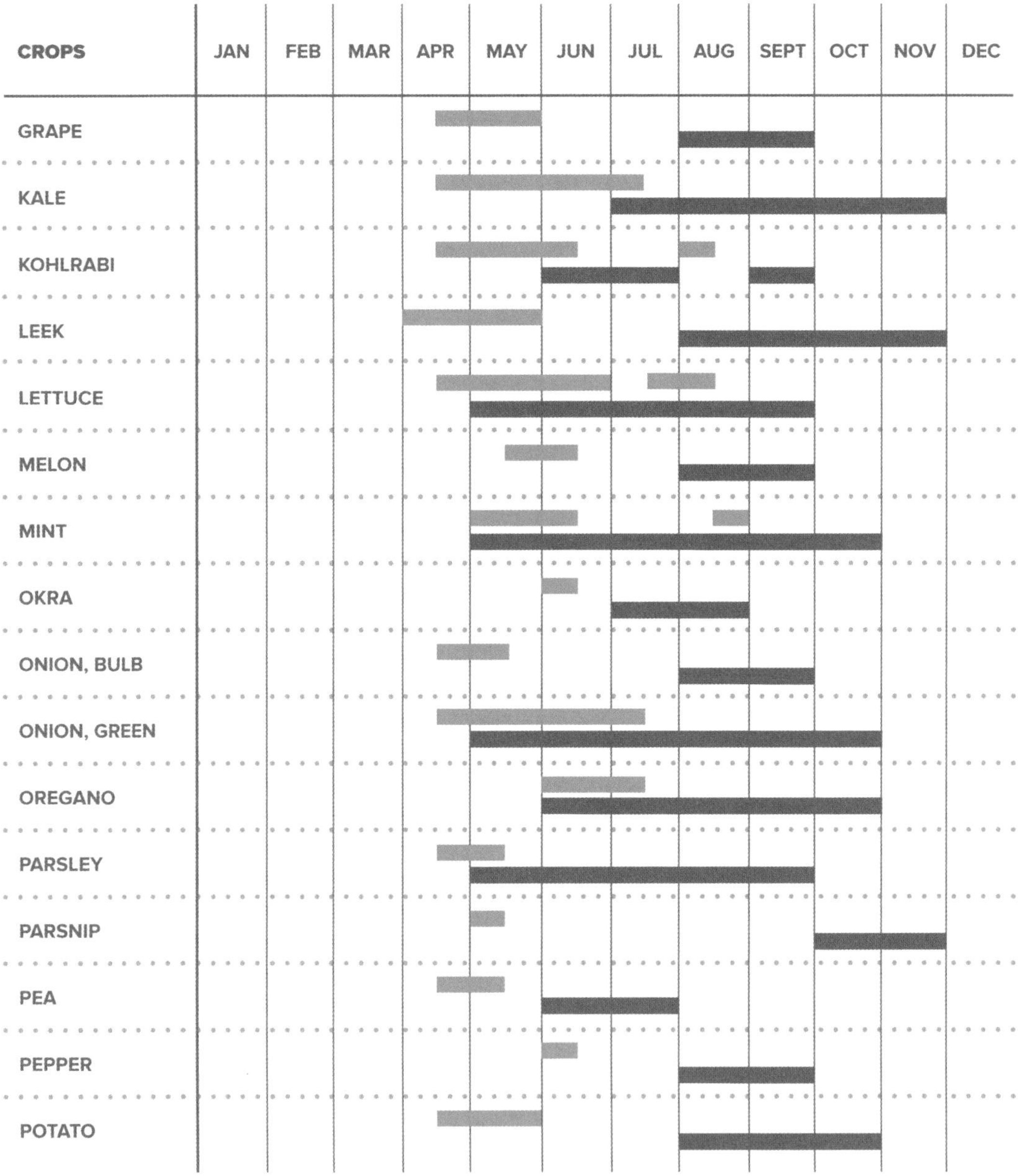

CROPS	JAN	FEB	MAR	APR	MAY	JUN	JUL	AUG	SEPT	OCT	NOV	DEC
GRAPE												
KALE												
KOHLRABI												
LEEK												
LETTUCE												
MELON												
MINT												
OKRA												
ONION, BULB												
ONION, GREEN												
OREGANO												
PARSLEY												
PARSNIP												
PEA												
PEPPER												
POTATO												

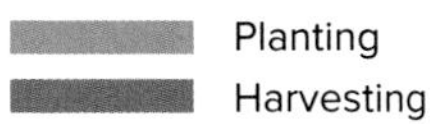

CROPS	JAN	FEB	MAR	APR	MAY	JUN	JUL	AUG	SEPT	OCT	NOV	DEC
PUMPKIN												
RADISH												
RASPBERRY												
RHUBARB												
RUTABAGA												
SHALLOT												
SPINACH												
STRAWBERRY												
SQUASH, SUMMER												
SQUASH, WINTER												
SWEET POTATO												
SWISS CHARD												
THYME												
TOMATO												
TURNIP												

PLANTING AND HARVESTING CHART

ZONE 5 AND 6

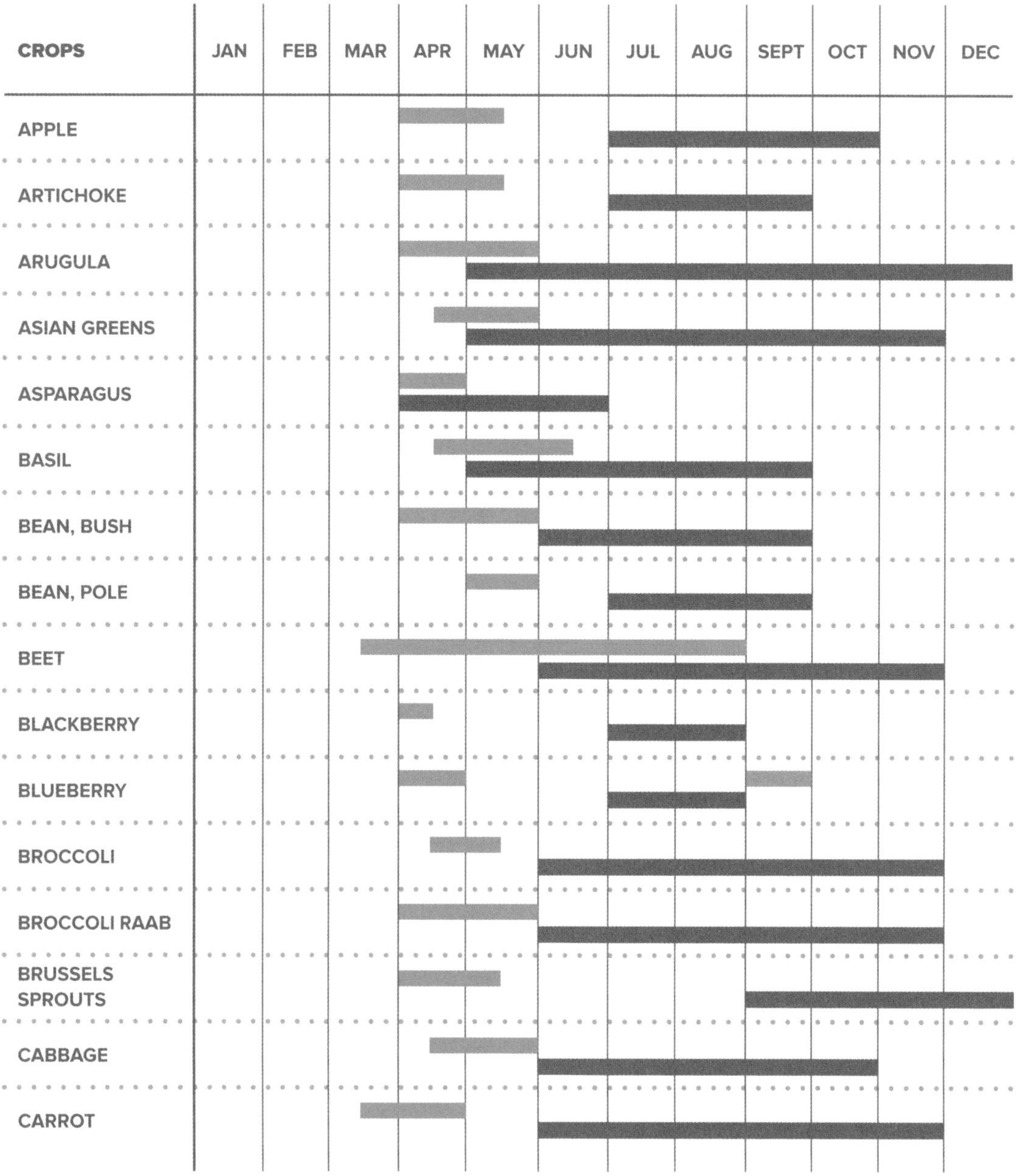

CROPS	JAN	FEB	MAR	APR	MAY	JUN	JUL	AUG	SEPT	OCT	NOV	DEC
APPLE												
ARTICHOKE												
ARUGULA												
ASIAN GREENS												
ASPARAGUS												
BASIL												
BEAN, BUSH												
BEAN, POLE												
BEET												
BLACKBERRY												
BLUEBERRY												
BROCCOLI												
BROCCOLI RAAB												
BRUSSELS SPROUTS												
CABBAGE												
CARROT												

CROPS	JAN	FEB	MAR	APR	MAY	JUN	JUL	AUG	SEPT	OCT	NOV	DEC
CAULIFLOWER												
CELERY/CELERIAC												
CHERRY												
CHICORY												
CILANTRO												
CORN												
CUCUMBER												
CURRANT												
EGGPLANT												
ENDIVE												
FAVA BEAN												
FENNEL												
GARLIC												
GOOSEBERRY												
GRAPE												
KALE												

CONTINUED ▸

PLANTING AND HARVESTING CHART

ZONE 5 AND 6, CONTINUED

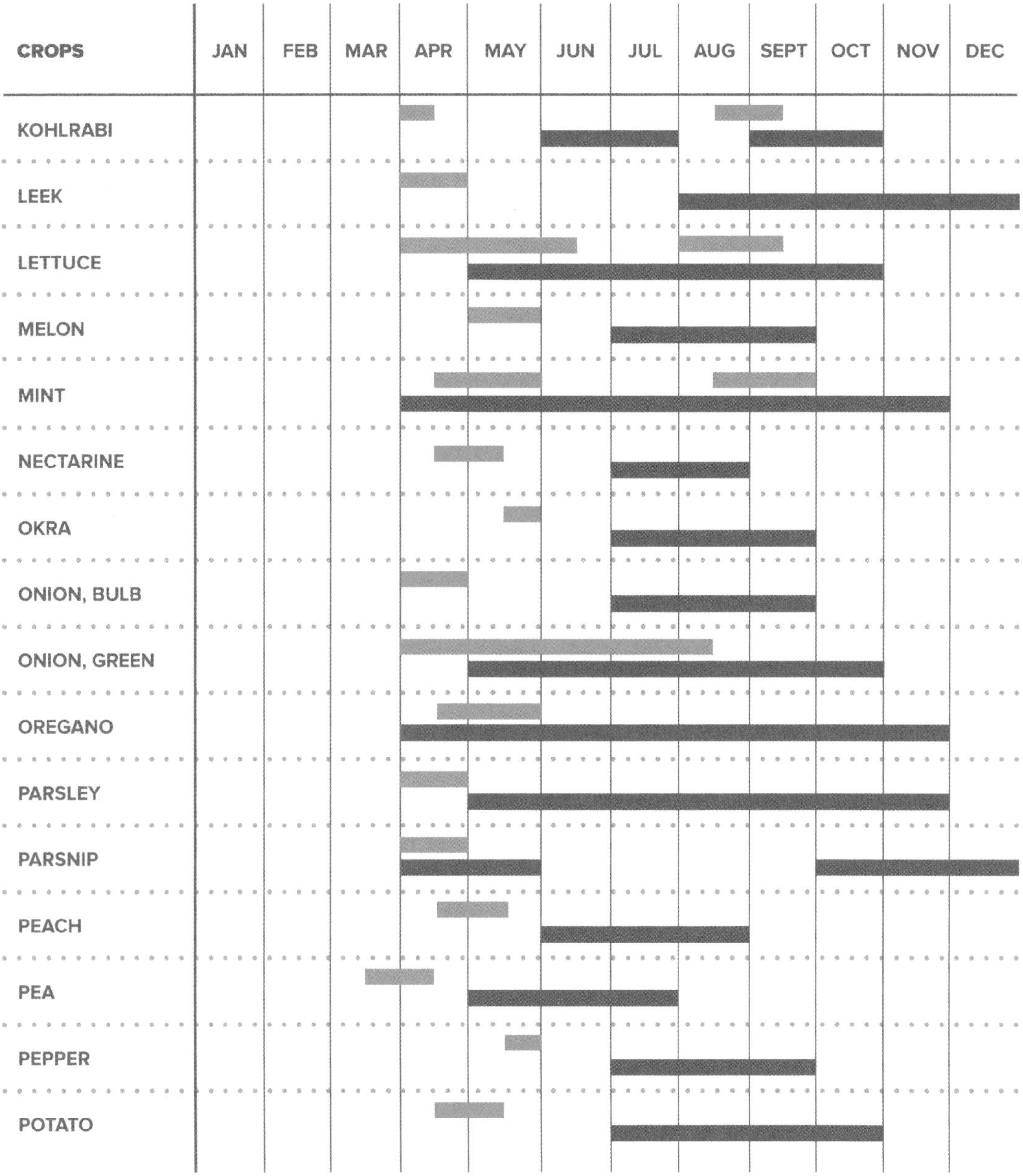

CROPS	JAN	FEB	MAR	APR	MAY	JUN	JUL	AUG	SEPT	OCT	NOV	DEC
KOHLRABI												
LEEK												
LETTUCE												
MELON												
MINT												
NECTARINE												
OKRA												
ONION, BULB												
ONION, GREEN												
OREGANO												
PARSLEY												
PARSNIP												
PEACH												
PEA												
PEPPER												
POTATO												

CROPS	JAN	FEB	MAR	APR	MAY	JUN	JUL	AUG	SEPT	OCT	NOV	DEC
PUMPKIN												
RADISH												
RASPBERRY												
RHUBARB												
RUTABAGA												
SHALLOT												
SPINACH												
STRAWBERRY												
SQUASH, SUMMER												
SQUASH, WINTER												
SWEET POTATO												
SWISS CHARD												
THYME												
TOMATO												
TURNIP												

PLANTING AND HARVESTING CHART

ZONE 7

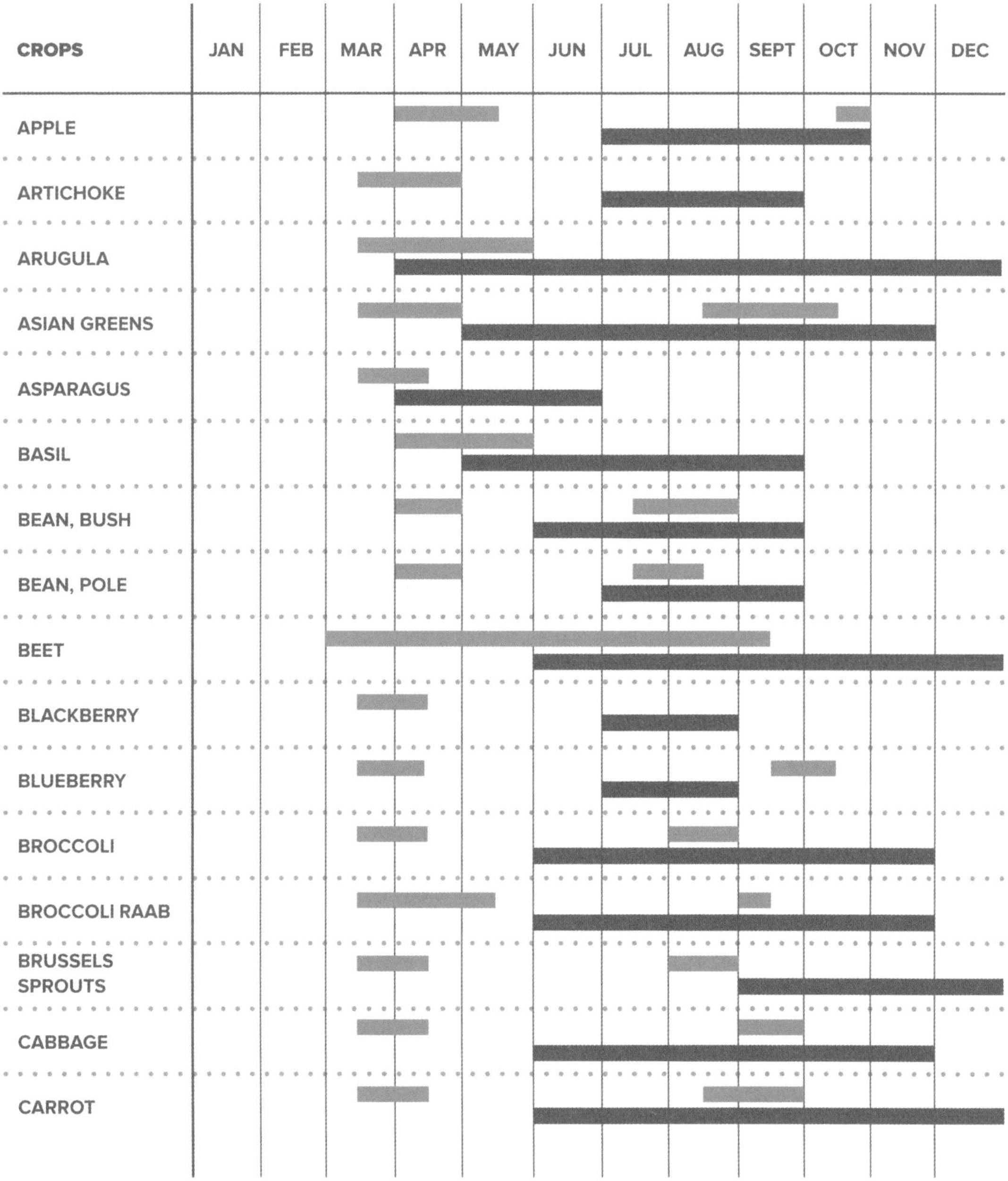

CROPS	JAN	FEB	MAR	APR	MAY	JUN	JUL	AUG	SEPT	OCT	NOV	DEC
APPLE												
ARTICHOKE												
ARUGULA												
ASIAN GREENS												
ASPARAGUS												
BASIL												
BEAN, BUSH												
BEAN, POLE												
BEET												
BLACKBERRY												
BLUEBERRY												
BROCCOLI												
BROCCOLI RAAB												
BRUSSELS SPROUTS												
CABBAGE												
CARROT												

CROPS	JAN	FEB	MAR	APR	MAY	JUN	JUL	AUG	SEPT	OCT	NOV	DEC
CAULIFLOWER												
CELERY/CELERIAC												
CHERRY												
CHICORY												
CILANTRO												
CORN												
CUCUMBER												
CURRANT												
EGGPLANT												
ENDIVE												
FAVA BEAN												
FENNEL												
GARLIC												
GOOSEBERRY												
GRAPE												
KALE												

CONTINUED ▸

PLANTING AND HARVESTING CHART

ZONE 7, CONTINUED

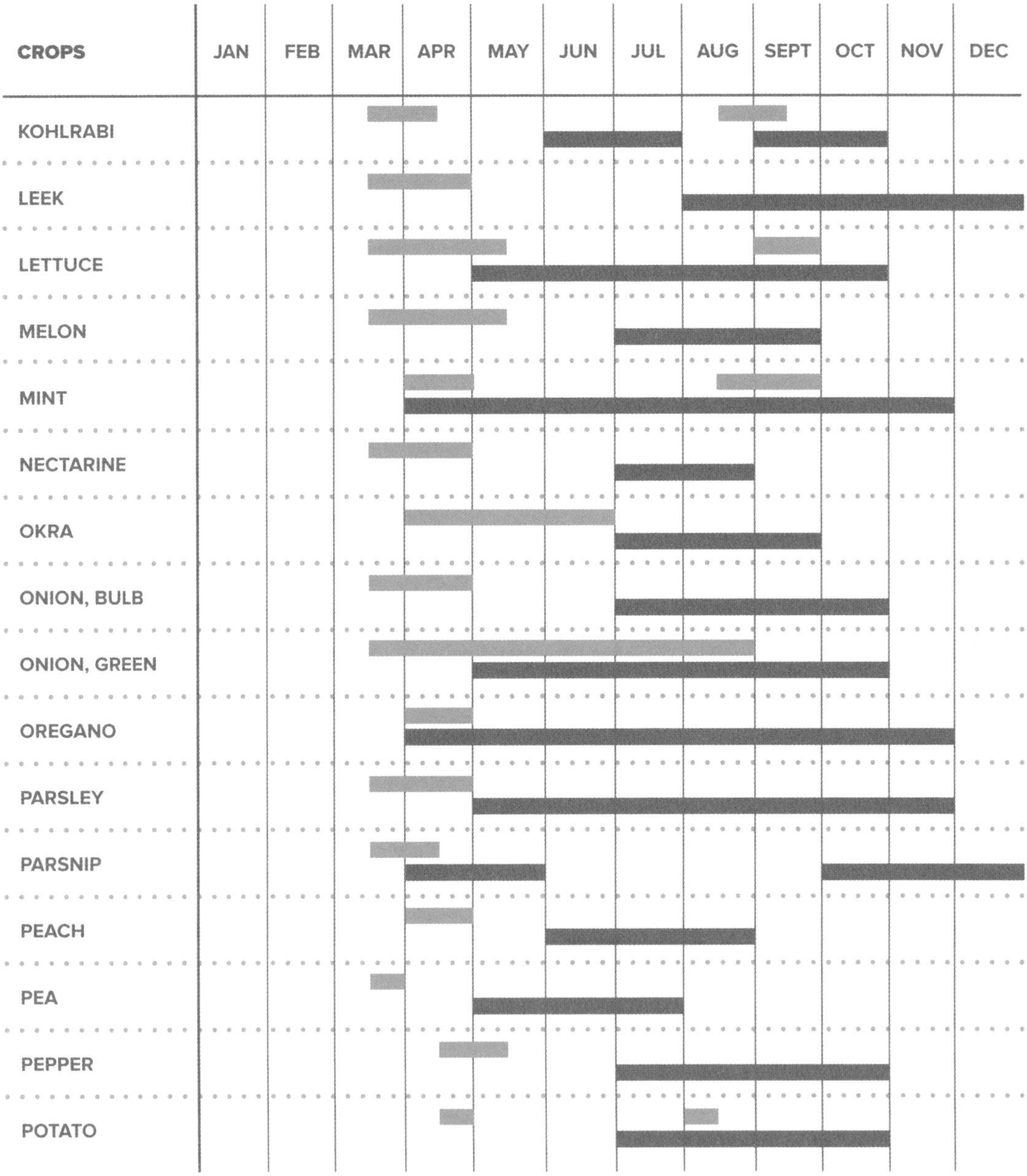

CROPS	JAN	FEB	MAR	APR	MAY	JUN	JUL	AUG	SEPT	OCT	NOV	DEC
KOHLRABI												
LEEK												
LETTUCE												
MELON												
MINT												
NECTARINE												
OKRA												
ONION, BULB												
ONION, GREEN												
OREGANO												
PARSLEY												
PARSNIP												
PEACH												
PEA												
PEPPER												
POTATO												

CROPS	JAN	FEB	MAR	APR	MAY	JUN	JUL	AUG	SEPT	OCT	NOV	DEC
PUMPKIN												
RADISH												
RASPBERRY												
RHUBARB												
RUTABAGA												
SHALLOT												
SPINACH												
STRAWBERRY												
SQUASH, SUMMER												
SQUASH, WINTER												
SWEET POTATO												
SWISS CHARD												
THYME												
TOMATO												
TURNIP												

Artichokes

New York is not often associated with growing artichokes. The conditions might not be ideal, but don't let that stop you from planting them. Fresh artichokes are more tender and flavorful than the seasonal offerings that pop up in grocery stores around Thanksgiving. They will require a little extra finesse and some winter protection if you want to grow them as perennials, but they are handsome plants—and what a treat it is to harvest your own baby artichokes to sauté.

GROWING Artichokes have a long growing season. Start seeds indoors in February or March. They germinate best in a warm (70°F) location. They are tender perennial plants that need a period of exposure to cool, but not freezing, weather—this is known as vernalization. To complicate things further, artichokes tend not to bloom until their second year of growth. To get around this, you can start plants from dormant roots, grow varieties that have been specially bred to set buds their first year, or fool them into thinking they've already gone through a winter and are starting their second growing season.

To fool first-year seedlings, move them to a cold but protected area about 2 weeks before your last frost date. A cold frame or tunnel is perfect. Keep them at 40 to 50°F for 3 to 6 weeks. After this little shock to their systems, you can transplant them to a sunny spot in the garden, where they will hopefully go to work setting tasty buds.

Artichokes are heavy feeders. Start with a rich soil and some supplemental organic fertilizer. Side dress with compost midseason. Plenty of sunshine and at least 1 inch of water per week will keep them growing. These are large plants, so space them at least 4 feet apart.

To overwinter established artichoke plants, you can either give them some extra protection with mulch or pot them up and bring them indoors. Bringing them inside is the safest option for colder climates, but this is a bit cumbersome with large plants. If you decide to bring them in, cut off the top of the plants, leaving 10 to 12 inches. Keep plants in a cool spot that receives some sunlight, and water lightly when the soil feels dry. Replant in the garden in May.

HARVESTING We eat the artichoke flower buds, and they should start forming sometime in July. The plants send up stalks with one central bud. Harvest this while the bracts are still tight but the bud feels dense. Don't wait until the flower starts to open or it will be tough. Slice the stem a couple of inches below the bud. Side shoots will form and more buds will follow.

VARIETIES **'Green Globe'** (90 to 100 days) is probably the hardiest variety. **'Imperial Star'** (80 to 90 days) was bred to set buds in its first year, but I expose it to an early chill. It starts producing early and is reliable. **'Opera'** (90 to 100 days), a purple hybrid, is very prolific. **'Violetto de Chioggia'** (80 to 90 days) has a short season for an Italian heirloom variety and is good for northern areas. **'Cardoons'** (60 to 65 days) are similar to artichokes in flavor, but you eat the stems, not the buds. With a short growing season and no need for vernalization, they are a tempting alternative.

Arugula

This cool-season green, with a fresh, peppery punch tempered by a sweet nuttiness, can dress up an everyday salad or stand on its own. You can eat the loose rosette of leaves fresh or lightly cooked, like spinach. Toss it into stir-fries, soups, and stews at the last minute, or add it to your favorite egg dish for some zip. It's a quick grower, and we can get several harvests in both spring and fall. During the hot summer, arugula turns bitter and bolts to seed.

GROWING Direct sow your first batch of arugula a couple of weeks before your last frost date. Broadcast the seed and cover with about ¼ inch of soil. Soil rich in organic matter will help it grow lush and green. Since it's a leafy green, arugula needs plenty of water, which shouldn't be a problem in spring and fall, although zone 7 gardeners may have better luck growing arugula in the fall. Succession sowing every 2 or 3 weeks will guarantee a long harvest season.

HARVESTING Begin harvesting by thinning plants spaced too closely together. Then start harvesting the outer leaves as they reach 3 or 4 inches tall. Arugula is a cut-and-come-again green; as the outer leaves are picked, the inner leaves fill in, and you can get several harvests this way. Or you can slice off an entire rosette.

VARIETIES Garden arugula (*Eruca sativa*) is often sold without a variety name, as simply arugula, rocket, or roquette. There's more variety to arugula than you might imagine, however. **'Apollo'** (45 days) has large, round leaves and a distinct sweetness. **'Astro'** (35 days) has slender, mild-flavored leaves. **'Roquette'** (35 to 40 days)—with a capital R—has wide, tangy leaves that can be hardy enough to overwinter. Wild arugula (*Diplotaxis muralis*) is an entirely different plant. This perennial can continue growing during the hot summer without bolting. The leaves are smaller than the annual varieties and it has a more intense peppery bite. You'll usually find it with some variation of the name **'Sylvetta'**.

Asian Greens

Asian greens run the flavor gamut from spicy to buttery. Some are enjoyed for their leafy greens, some for their crunchy ribs, and many even have edible flowers and buds. Most prefer the short, cool days of fall, but you can easily get a spring harvest, too. They are quick growers and do not require much space, so you'll be able to experiment with the dozens of varieties to find your favorites.

GROWING Sow in early spring and again in late summer, about 8 to 12 weeks before your first fall frost. You can also start seedlings indoors to transplant into a moist, fertile soil in a sunny location. Sow about ¼ inch deep and keep well watered. The seeds germinate quickly, within a week. Thin seedlings when they are 1 or 2 inches tall, and eat them fresh or cooked. For loose-leaf greens, thin to about 6 inches. Heading varieties should be spaced up to 12 inches apart.

HARVESTING Leafy greens can be harvested as a few leaves or in whole clumps. You can pick them as baby greens or allow them to mature into heads. If you cut the entire clump and leave the crown intact, the plant will resprout several times. Heading varieties can be cut whenever they reach their mature size. Most Asian greens can withstand a slight frost, but harvest them before a hard freeze.

VARIETIES The mustards (mizuna, tatsoi, red mustards) grow fast and pest-free. Cool weather makes them sweeter. Japanese chrysanthemum (*Chrysanthemum coronarium*), sometimes called chop suey greens (20 to 60 days), has a pleasant bitterness and spicy flower petals. **'Extra Dwarf'** pak choy (30 to 45 days) matures at 2 to 3 inches and is piquant and tender. Choy sum (35 to 50 days), or choi sum, is a flowering mustard that produces lots of small, spicy-sweet edible flowers.

Asparagus

Growing asparagus requires some initial patience, but then you get to sit back and indulge. This perennial crop produces for 20 years or more, so find a spot where it can grow undisturbed. Each spring you'll have multiple harvests of pencil-thin, sweet spears.

GROWING Although you can grow asparagus from seed, doing so will add a year or two to your wait time. Starting with 1-year-old crowns is just as easy and almost as inexpensive. The bare-root crowns look like badly worn string mops.

Prepare the bed in early spring. Dig a trench 8 to 10 inches deep and 18 to 20 inches wide. Work some compost into the bottom of the trench and amend your soil to a slightly acidic pH 6.5. Spread out the roots of the crowns on the bottom of the trench, spacing plants 12 to 15 inches apart. Cover with 2 or 3 inches of soil and water well. As the plants grow, continue covering them with soil until the trench is full.

Young asparagus will start out as the familiar spears. As the plants mature, they burst into airy fronds. Keep the beds weed free and top dress each spring with a fresh layer of compost or rotted manure. I prefer to cut back asparagus in the fall to prevent overwintering problems (such as hiding insects), but you can leave them in place as mulch if you wish. Just be sure to cut them back in early spring, before new growth starts.

Asparagus has few problems, but watch for asparagus beetles, which can feed on the fronds and weaken the plants. Hand pick or use an organic pesticide, such as neem, if necessary.

HARVESTING Don't harvest anything the first year; let the plants grow and build up strength for the long haul. The second year, you can harvest a few thicker spears, but let them grow undisturbed after that. By the third year, you can harvest as much as you want. When the appearance of new shoots begins to slow, you'll know it's time to let the plants grow.

Harvest when spears reach 6 to 8 inches tall; snap them off at ground level. You can also cut them at the soil line, but take care not to slice through any emerging spears. When the plants hit their stride, you'll be harvesting for up to 2 months.

VARIETIES **'Mary Washington'** (720 days) is an heirloom variety that is particularly suited to the Northeast. Newer varieties are all male and considered by some as better producers. **'Jersey Knight'** and **'Jersey Supreme'** (both 720 days) are both hardy from zones 3 to 8. **'Purple Passion'** (720 days) is especially tender; the purple color disappears with cooking.

Beans

Beans are one of the easiest vegetables to grow, and they come in dozens and dozens of varieties and types. They can grow on bushes or vines, can ripen in early season or summer, and can be snapped, shelled, or dried. The hard part is limiting yourself to a handful of varieties.

Bush beans are short plants that begin producing earlier than the pole varieties. They produce for 4 to 6 weeks and can be succession planted. Pole varieties take longer to develop pods, but once they do, you'll harvest them continually until the end of the season.

GROWING Beans have big, easily handled seeds. Wait until the soil has warmed in the spring to about 60°F and is dry enough that you cannot form it into a ball. Planting seeds in cool, wet soil will cause them to rot. Many gardeners like to use inoculants for the first spring planting, which helps protect the seed from late-season chills and rains.

Pick a sunny spot and plant the seed 1 inch deep. Space seeds 2 to 4 inches apart or broadcast them in blocks. To keep bush beans producing, succession plant every 3 to 4 weeks. Pole beans will need to climb on a support or trellis. Sow four to six seeds at the base of each pole of a teepee or space them 2 to 4 inches apart on a trellis.

Keep the plants watered, especially while they are in bloom. They should not need supplemental food, although pole beans could do with a side dressing of compost in midsummer. Cut old bush bean plants at the soil line when they finish producing. Leave the roots in the soil, so that the attached nitrogen nodules can decompose and improve the soil.

HARVESTING Snap beans are ready to harvest when the pods feel firm and are just starting to fill out. Don't wait until you can see the seed shape through the pod; young beans are the most tender and flavorful. Shelling beans are mature snap beans. Wait until the beans plump up and look lumpy, and then pick and pop the beans out of the pods. For dried beans, let the pods dry on the vine before picking and shelling.

VARIETIES **'Kentucky Wonder'** (60 to 70 days), **'Blue Lake'**, and **'Kentucky Blue'** (both 55 to 65 days) are popular standards available in both bush and pole varieties. **'Lazy Housewife'** (75 to 80 days) is an heirloom that produces prodigious clusters of beans for easy picking. **'Cherokee Wax'** (50 to 55 days) is a juicy, stringless wax bush bean. **'Black Valentine'** (50 to 55 days) can be grown as a snap or shelling bean and grows well in cool soil. **'Cannellini'** (75 to 80 days shell, 95 to 100 days dried) is a creamy, white Tuscan bean.

Beets

Few vegetables are as versatile as beets. You can eat the thinned plants and the tender tops, both with an earthy savoriness and just a hint of sweetness. The bulbs are great raw when grated into salads and soups, roasted until their sugars caramelize, steamed and sautéed, and even pickled. They come in beet red, golden orbs, and peppermint spirals. All this and they can grow pretty much all season long.

GROWING Beets are easy to squeeze into the garden. You can underplant late-maturing vegetables with beets or give the beets a block or row of their own. They prefer full sun and a rich soil. Beet seeds come in clusters. If you plant the whole cluster, you'll need to thin out the tiny seedlings to 2 or 3 inches apart; thinnings make fine eating. You can also gently crush the clusters and plant the individual seeds. Direct sow after the soil has warmed to about 50°F. Soaking the seeds overnight before planting will soften their tough shells and speed germination. Keep the plants well watered, but allow the soil to dry between watering. The leaves will form a natural mulch. To keep the harvest coming, succession plant every 3 or 4 weeks through midsummer. Beets slow down in the heat of summer, but you can resume planting in the autumn after nighttime temperatures fall to about 75°F.

HARVESTING Beets are sweetest and their most tender when harvested small, at 2 to 3 inches. You can harvest some of the greens while they are young and tender; they taste similar to Swiss chard, a close cousin.

VARIETIES **'Chioggia'**, or bull's eye beet (60 days), is as beautiful as it is sweet. **'Touchstone Gold'** (55 days) is sweeter still. **'Detroit Dark Red'** (60 days) forms perfect round balls that stay tasty in storage. **'Bull's Blood'** (60 days) is grown for its succulent, blood-red leaves.

Blueberries

Berries are perfect candidates for edible landscaping, and blueberries, one of the few berries native to North America, will give you three seasons of beauty and interest in the garden. These perennial shrubs will keep you in pints of sweet-tart berries for years, with minimal maintenance and pruning required. Small lowbush or wild blueberries (*Vaccinium angustifolium*) mature in July and are considered the most flavorful by many. They are grown mostly in Maine and parts of Canada and are hardy in zones 4 to 8. Cultivated highbush blueberries (*V. corymbosum*) are larger and mature earlier than lowbush berries. They are the most commonly grown varieties and are hardy in zones 3 to 9. Half-high blueberries are being developed, which combine the larger size with the flavor of lowbush. Rabbiteye blueberries (*V. ashei*) are the tallest bushes, with prolific fruits similar to highbush. They are not terribly cold hardy, but they are very drought tolerant and good choices for zones 7 to 9.

GROWING Blueberries absolutely need acidic soil, in the pH 4.5 range. If your soil isn't naturally acidic, amend it with garden sulfur or aluminum sulfur in the fall before you plant.

Although some blueberry plants are self-fertile, you'll get more berries if you plant two varieties that bloom at the same time, to cross pollinate. Younger plants at 2 or 3 years old transplant best. Plant in early to midspring, spacing bushes about 4 to 5 feet apart. Water them well and keep them watered all season. Blueberries are shallow rooted and need several inches of water each week. Do not fertilize the plants the first year you plant them. In subsequent years, feed with a fertilizer labeled for acid-loving, edible plants.

Remove all the flowers from your blueberry plants during their first season. They will start to produce berries in their second year but won't be fully mature and producing high yields until their fifth or sixth year.

HARVESTING Blueberries are ripe when they are uniformly blue with a white, powdery dusting. They'll pass through shades of green and lavender first. As they start to develop their ripe color, consider netting them or hanging old CDs or strips of aluminum foil to keep the birds away. Early-season fruits ripen in June and July. Midseason berries ripen in late July, and late-season berries are ready in August. As always with fruit, the surest way to know when your berries are sweet enough to pick is to sample a few.

VARIETIES To prolong your blueberry pleasure, plant early-, midseason, and late-season varieties. Highbush varieties to try include early-season **'Earliblue'** and **'Collins'**, midseason **'Blueray'** and **'Bluecrop'**, and late-season **'Jersey'** and **'Elliot'**. Lowbush varieties include **'Brunswick'**, **'Burgundy'**, and **'Top Hat'**. Rabbiteye varieties include **'Tifblue'**, the old standard. For early-season varieties, try **'Climax'** and **'Woodard'**; for midseason, try **'Briteblue'** and **'Southland'**; and for late-season, try **'Delite'**. New pink-berried varieties include **'Pink Lemonade'**. Several blueberries are suitable for container growing, including midseason **'Dwarf Northblue'** and late-season **'Dwarf Tophat'**.

Broccoli

Broccoli is actually an edible flower, and we eat the giant cluster of flower buds—delicious flower buds. It is not the easiest vegetable to grow well, partly because it needs cool but not cold weather and partly because vegetables in the brassica family (such as cabbage, cauliflower, and kale) are favorites of many pests. So why grow it? Homegrown broccoli has a good deal more flavor and tenderness than the limp heads sold in the grocery store, and you can grow delightful varieties that you won't find in the produce aisle.

GROWING You can direct sow broccoli about a month before your last frost date, but young plants are sensitive to cold. If they get hit by frosty temperatures, they'll never produce a head, so be prepared to cover and protect them. You're better off starting with transplants or starting seeds indoors and moving the plants out after danger of frost is past. The fall means fewer pests and long, cool days; this is an excellent season for growing a crop of broccoli, but you will need to start the plants in midsummer so they'll be ready to grow on in fall.

Broccoli is a large plant that will grow tall first and then reach out with its leaves. Sow seed ¼ inch deep, spaced 6 inches apart, and thin the plants to 12 to 24 inches. Give them a rich soil and plenty of water and sunshine.

Brassicas are magnets for a parade of insect pests. To prevent insects from laying eggs on your plants, grow broccoli under row covers. Otherwise, keep a keen eye out, searching for and destroying eggs that are often on the undersides of leaves.

HARVESTING The central head is ready to cut when it is full and firm but the buds are still tightly closed. Don't wait too long or the yellow flowers will burst open; you can still eat them, but they won't taste as good as they should. Use a knife to harvest only the head and a short bit of stem; leave the rest of the plant intact, and smaller side shoots will continue to form. The more you harvest, the more broccoli you will get.

VARIETIES Mix things up with varieties that mature at different rates. **'Packman'** (50 to 55 days) gets off to a quick, dependable start. The head is small but side shoots continue for months. **'Piracicaba'** (55 to 60 days), pronounced "peer-a-Ceeca-bah," has an impressively long sprouting season and stands up well in both heat and cold. **'Premium Crop'** (55 to 65 days) forms a large head and many side shoots. **'Purple Sprouting'** (200 days) is meant to be planted in late summer and overwintered. Come spring, you will be harvesting an abundance of tender shoots. **'Romanesco'** (75 to 80 days) forms stunning chartreuse, spiraling, conical florets that arrange themselves in a fractal pattern.

CABBAGE FAMILY PROBLEMS

The members of the brassica family, which includes broccoli, cabbage, cauliflower, collards, kale, and mustard, all tend to suffer from the same diseases and pest problems. Many of these pests and diseases can overwinter in the soil. To avoid problems, clean out all garden debris in the fall and plant brassicas in a different section of the garden each year.

FUNGAL DISEASES

Fungal diseases are spread by spores and can rapidly infect nearby plants. They are usually exacerbated by damp weather. The best defense is to choose resistant varieties, keep the growing area clear of infected debris, and rotate where you plant brassicas every year.

Blackleg. Look for dark, sunken areas on the stem; plants also wilt. Damp weather will encourage the disease to spread.

Clubroot. This soil-borne slime mold makes the roots swell, causing yellowing, wilting, and stunting. The good news is that the mold does not like alkaline soil; adding enough lime to raise the soil pH above 7.2 can help control it.

Fusarium yellows. Plants, primarily cabbages, turn yellowish shortly after planting and get worse during hot, damp weather. The condition may improve as the weather gets drier. Look for varieties labeled as resistant to YR.

White mold. Also known as Sclerotinia, this fungus generally affects late-season cabbages. It starts as tan, water-soaked spots that eventually become covered in fluffy, white fungus. No fungicides are approved for white mold control, so you should destroy the affected plants (do not compost them because the fungus spreads easily) and rotate your planting area.

INSECTS

Brassica insect pests are early risers and are some of the first pests to arrive in the spring. Because brassicas are in the garden while it is still cool outside, they are the food of choice. The best defense here, other than careful monitoring and hand picking, is to protect early seedlings with row covers.

Cabbage aphid. Like its aphid cousins, these insects suck plant juices from the undersides of the leaves, transmitting diseases as they feed. Leaves are stunted and puckered. Remove the affected leaves and spray the whole plant with insecticidal soap.

Cabbage looper. This common, pale green worm feeds mostly on the undersides of leaves. Watch for multiple generations each season, and hand pick or use *Bacillus thuringiensis* (Bt).

Cabbage maggot. This small white larva bores its way into stems and roots, killing the plants in the process. You can place paper collars around the base of the seedling stems to block the insects' entry. Row covers will prevent the adult fly from laying eggs near the plants.

Cutworm. Another moth larva, cutworms are most infamous for chewing right through young stems at ground level. Add paper or foil collars around the stem to prevent cutworms from circling the plant and cutting through. Or stick toothpicks halfway into the ground next to the stem. Place them on either side of the stem, being careful not to stab into it.

Imported cabbage worm. That pretty, pale yellow butterfly that hovers around your cabbage and kale is laying eggs that will hatch as the dreaded cabbage worm. Larvae are green with dark stripes and turn leaves into Swiss cheese as they feed. They tend to leave telltale black droppings. You can hand pick or add row covers, use Bt, or sprinkle diatomaceous earth or wood ashes around the plant to control these pests.

BACTERIAL DISEASES

Black rot. This bacteria gets inside cabbage and cauliflower plants, causing leaf veins to blacken. Heads may rot or never form at all.

Broccoli Raab

Broccoli raab, or rapini, looks like a tiny broccoli plant, but it is actually a closer relative of the turnip. The flavor is similar to that of broccoli, and you can use it in much the same way, except that you eat the entire plant: the leaves, stems, and small flower buds. This is another quick-growing, cool-season plant that thrives in the drizzly days of spring and fall.

GROWING Direct sow in early spring, as soon as the soil can be worked. Broadcast the seeds and cover with about ¼ inch of soil. Start thinning when the plants are 3 inches tall, and toss the thinnings in any dish. Mature plants will reach about 6 to 8 inches tall.

Succession plant every 3 or 4 weeks throughout the spring. Broccoli raab does not do well in summer heat, but you can resume sowing in late August or September for a fall harvest. Fall sowings do better in a semi-shady spot, perhaps under a trellis; cool the soil with a good soaking. Keep the plants well watered and be on the alert for slugs that will want to share the damp shade as they gobble up your plants.

HARVESTING After thinning, your first official harvest should begin when the flower buds start to form but before they open. As with broccoli, leave 4 or 5 inches of the stem on the plant to resprout a few times.

VARIETIES You'll often find seed labeled only as broccoli raab or rapini, but there is some variance in flavor and growth among the varieties. Look for and experiment with quick-growing and tender **'Quarantina'** or **'Sorrento'** (both 40 days); **'Spring Raab'** (45 days), which has a sweeter bite to it; and **'Marzatica'** (75 to 110 days), which is best for fall and overwintering in milder areas.

Brussels Sprouts

This cabbage cousin creates an exotic look in the garden and is a welcome sight on the plate. The sprouts form along a tall stem and ripen from the bottom of the plant up. They need a long growing season and the caress of frost to make them sweet and succulent. Find a spot in the garden where they can spend the summer, and then be patient.

GROWING Because Brussels sprouts take months to mature, they are usually started from transplants rather than directly sown. You'll have more control that way, too, and another month without them hogging precious garden space. They prefer rich, sweet soil with a pH of at least 6.5.

Set out your transplants after the threat of a hard frost. Gardeners in zone 7 should hold off planting until midsummer, when seeds can be sown for a late fall harvest. To start indoors, cover the seed with ¼ inch of soil and keep them well watered. When seedlings are about 3 inches tall, you can transplant them into the garden, spaced 2 or 3 feet apart: these are large plants. You can take advantage of the empty spaces between plants early in the season by planting a quick grower such as lettuce or spinach. Keep the plants watered and side dress them twice a season, about a month after planting and a month before harvest.

Unfortunately, Brussels sprouts are susceptible to the same pests that harass the rest of the cabbage family, but because they start setting their sprouts late in the season, you can usually get any problems under control before they affect the sprouts.

HARVESTING A couple of light frosts should be enough to sweeten the sprouts. They are ready to pick when they are about 1 to 1½ inches in diameter and feel full and firm. Because they mature from the bottom of the plant up, start harvesting the lower sprouts and let the rest fill in. You may get a second flush of smaller sprouts at the bottom of the stem.

Brussels sprouts can tolerate freezing temperatures and even a little snow. In colder areas, protect the plants with row covers or straw mulch and harvest into winter—but harvest them before they get so large they crack, because they'll be bitter by then. You can speed up the development of sprouts near the end of the season by cutting the plants' leafy tops, which are delicious when cooked. To keep your Brussels sprouts fresh longer, cut the whole stem for your last harvest and don't remove the sprouts until you are ready to cook them.

VARIETIES Try **'Bubbles'** (85 to 90 days), which has tender, sweet sprouts up to 2 inches and handles heat and drought well. **'Long Island Improved'** (90 days) is a smallish plant with a high yield. **'Rubine'** (85 to 95 days), a pretty reddish plant, matures later and has a lower yield than the early green varieties but a delicious flavor.

Cabbage

A growing head of cabbage is an imposing sight. The expanding, tight center ball is surrounded by huge floppy leaves, which makes the plant look much like a cabbage rose. Some cabbages can take up a lot of garden real estate, but others are not so large. Some good dwarf varieties have been introduced especially for home gardens, and crinkly Napa cabbage has a more upright growth habit, requiring less room.

GROWING Cabbage likes chilly weather, and you can set seedlings out 2 or 3 weeks before your last frost date. If you are starting seed indoors, sow seed 4 to 6 weeks before you plan to set them out. Gardeners in warmer areas can direct sow after danger of frost. We can all get a second season in by planting again in midsummer to late summer, about 12 weeks before the fall frost date, to mature in the fall. Fall-grown cabbages are less prone to splitting.

Cabbages need a rich, sweet soil. If you are direct sowing, space seeds about 4 inches apart and cover with ½ inch of soil. Final spacing will depend on the variety you are growing, but each full-sized head will need 2 or 3 feet of space. Consistent watering is the key to getting a plump, firm head. Cabbages attract the usual brassica pests, with cabbage worms topping the list.

HARVESTING Harvest when the heads feel firm. Wait too long and the heads will split open and won't store well. Leave the wide, outer leaves and slice off the head at its base. Cutting an X in the top of the remaining stalk sometimes results in four smaller heads forming, but I've never had much luck with that.

VARIETIES **'Fast Ball'** (45 days) lives up to its name and produces crisp, softball-sized heads. **'Early Jersey Wakefield'** (60 to 70 days) is a space-saver with a large pointed head and few outside leaves. **'Rubicon'** (F1) (50 to 55 days), a tight-headed Napa, is self-blanching and resists bolting. **'January King'** (100 to 120 days) is a gorgeous, purple-tinged semi-savoyed heirloom.

Carrots

Root vegetables always require a leap of faith to grow, because we don't know what's going on down there until we dig them up. Carrots can go wrong in so many ways: they can fork or otherwise become misshapen, be tough or bland, or be riddled with insect tunnels. But we love them anyway, and when we find a variety that works, a fresh carrot is sweet as nectar.

GROWING Carrots like moisture, cool weather, and loose soil. Hitting a rock or a hard clump is enough to cause a carrot to fork or deform. If you're concerned about your garden soil, consider growing carrots in a raised bed or container. Another carrot caveat is soil pH: carrots will not grow well in very acidic soil, preferring a pH in the 6.5 to 6.8 range.

Begin direct sowing as soon as the ground can be worked, even if it is a few weeks before your last frost date. Succession plant every 3 or 4 weeks while the weather remains cool. Carrots need to grow fairly fast to become sweet. In zone 7, fall and late winter are the best sowing times. Sow the seeds about ¼ inch deep, spaced 1 or 2 inches apart. The seeds are tiny and hard to work with, so if you over-sow, you can thin the plants later and eat the thinnings. Carrots can be frustratingly slow to germinate, taking up to 2 weeks. Don't let the soil surface crust over, and keep the soil moist until you see the first seedlings poke through. Keep your carrots well watered the whole time they are growing, and add a layer of straw mulch to help keep the soil moist and cool.

The major carrot pest, other than the four-footed marauders, is the carrot rust fly. It lays its eggs twice a year at the base of the plants, and the larvae, or maggots, tunnel into the roots as they feed. The pupa stage overwinters in the soil, so moving your carrots to a new location every year or planting them only every other year will help keep the larvae at bay.

HARVESTING Check to see if your carrots have reached full size by feeling around for the width of the tops just below the soil surface. If they are the right size for the variety you are growing, dig up a couple; first loosen the soil with a fork, because pulling on the top often means pulling off the top. Fall carrots can remain in the ground through fall and winter and even harvested in winter if the ground is not frozen. Cool temperatures will sweeten them; however, if rust flies have been a problem, it's best to dig them all out at the end of the season.

VARIETIES **'Chantenay Red Cored'** (70 days) and other **'Chantenay'** and **'Nantes'** varieties mature at 4 to 5 inches and are good choices for rocky soils. **'Imperator'** (72 days) is a good choice for a longer, 7- to 9-inch, tapered carrot that stores well. Or try one of the unexpected colorful carrots, such as the purple **'Dragon'** (70 days) or the creamy **'Snow White'** (70 days). For baby carrots, try the **'Paris Market'** or **'Thumbelina'** (both 50 to 55 days) nugget-sized treats.

Cauliflower

Cauliflower knows what it likes and will not accept anything less; that makes it a little exacting to grow. Like all the brassicas, it prefers cool weather; too much heat or dry weather, and you can wind up with small button heads. Add too much nitrogen, and the curds (the florets) turn grainy or ricey. I have found the purple varieties easier to grow, because they do not require blanching (covering to keep them white).

GROWING Start seed indoors, 4 to 6 weeks before your last frost date. The seeds like warm soil, 70 to 75°F, but once they've germinated, keep the plants a cooler 60°F. Transplant outdoors in 4 to 6 weeks, spacing the plants 12 to 24 inches apart. Here's where the tricky part comes in, because if the weather fluctuates, the plants may never form heads. Wait until the soil has warmed to 50°F and there is no danger of frost. Using row covers over young plants helps moderate temperature swings, with the bonus of protecting transplants from insects. Cauliflower is susceptible to the same pests that affect broccoli and cabbage.

You can direct sow a fall crop in early summer or set out transplants in midsummer. Fall cauliflower often heads better, but be sure to keep plants well watered. The roots are very shallow and mulching is advised.

For truly white cauliflower, you'll need to blanch the heads by covering them with the plants' outer leaves, starting when the head is about 2 or 3 inches across.

HARVESTING Harvest by cutting the stem just under the head when the head feels firm and it's just beginning to form curds. You may get a second flush of smaller curds.

VARIETIES **'Snow Crown'** (50 to 60 days) is less temperamental than most varieties. The colorful varieties spare you the need for blanching. Try the aptly named **'Cheddar'** (80 to 100 days) or my favorite, the vibrant purple heirloom **'Di Sicilia Violetto'** (80 to 90 days).

Celery and Celeriac

Speaking of challenges, let's talk about celery, yet another cool-season plant that sulks and bolts in heat. It also takes an interminable amount of time to germinate, bolts if exposed to a prolonged cold spell while still a seedling, and requires blanching. However, some gardeners think the bracing flavor of fresh celery is worth the effort. I prefer growing celeriac, which has all the flavor and crunch of celery, except you eat the much easier to grow bulbous root.

GROWING Start seed indoors, 10 to 12 weeks before your area's last frost date. The seeds need light to germinate, so press them on the surface of the soil and do not cover them. Plant extra seeds, because the germination rate is poor.

Keep the soil at a warm 70 to 75°F and keep it moist, and then wait. Germination can take up to 3 weeks. When the seeds sprout, move them to a cooler 60 to 70°F spot, and thin to one plant per cell. Transplant the seedlings outdoors about 2 weeks before your last frost date. Young plants can handle a little frost. You can space both celery and celeriac fairly closely, about 6 to 8 inches apart. Mulch the celeriac to keep the bulbs covered. Zone 7 gardeners can start a fall crop in midsummer, and it may even overwinter for you.

Celery doesn't have many disease problems, but insects can become pests. Aphids and whiteflies can weaken the plants and the tarnished plant bug can ruin the leaves, but cutworms can completely destroy young plants by cutting them off at ground level. Once again, row covers early in the season are your best line of defense.

HARVESTING Harvest celery when the bunch is 2 or 3 inches thick and feels firm. The homely celeriac bulbs will push themselves up out of the soil. Harvest when the bulb is about 3 to 5 inches in diameter. Celeriac's root system is extensive, and you'll need to use a fork to lift out the bulb and roots. Use the bulb like celery in cooking. You can also cut off and use the roots and stalks, which are great for flavoring stocks.

VARIETIES *Celery*: **'Tall Utah'** (90 to 100 days) resists bolting and doesn't get stringy. **'Ventura'** (80 to 100 days) has large, flavorful leaves and handles frost well. *Celeriac*: **'Mars'** (90 days) has a bulb that stays firm, no matter how large it grows. **'Brilliant'** (100 days) has relatively smooth bulbs. **'Giant Prague'** (110 days) is a large and crispy heirloom.

Chicory and Endive

The names chicory, endive, escarole, and radicchio are tossed about interchangeably. They are all in the same family, and it can be difficult to keep them straight. Cutting chicories (*Cichorium intybus*) are biennial plants, and radicchio is a type of chicory. Most chicories grow into rosette heads, unlike endives, which are loose-leaf annuals. Roots are often forced (removed to a warm place to be coaxed into rapid and early growth), but they can be grown informally in the garden. Escarole (*C. endivia*) is a type of endive with flat leaves. Belgian endive (*C. intybus*) is the familiar small head sold at grocery stores.

GROWING These cool-season plants are grown for their leaves. Despite their fancy names, they like the same growing conditions that lettuce prefers: a rich, well-draining soil; mulch to keep the soil cool; and plenty of water.

Start chicory seed indoors 6 to 8 weeks before your last frost date. Transplant seedlings into the garden when they are 4 weeks old. Endive seed can be started 4 to 6 weeks before the last frost date. For fall crops, direct sow in midsummer. Give them regular water, and that's about it. Both plant families are extremely cold tolerant and will grow well into fall and milder winters.

HARVESTING Harvest when the heads reach a usable size. You can blanch plants in the garden, for a milder flavor, by covering them with a basket or something similar that lets air, but not light, get through. Both chicory and endive can be eaten fresh in salads or braised and grilled for a sweeter flavor.

VARIETIES *Chicory*: **'Bianca Riccia'** (35 to 40 days) forms a rosette of tender, serrated leaves. **'Catalogna'** (48 days) has tangy, strappy leaves and is slow to bolt. *Radicchio*: **'Rossa di Treviso Precoce'** (60 days) is a sweet heading variety that turns deep red in the fall.

Corn

There is no denying that freshly picked corn is so sweet and tender that it needs no cooking. But corn takes up a lot of space in the garden, and wonderful sweet corn is available at every farm stand throughout the Northeast. Many gardeners prefer to leave it to the farmers and grow crops that are not available elsewhere. I would encourage everyone to try growing corn at least once, however, for the sheer joy of the harvesting and eating experience.

GROWING Corn has a long taproot and is a heavy feeder. It needs a loose, nitrogen-rich soil, so provide plenty of organic matter. Direct sow seed after all danger of frost has past. Corn is pollinated by wind, which moves the pollen from the tassels to the silks. Pollination is more easily accomplished when stalks are planted closely in blocks rather than in rows. Sow seed 1 inch deep and 4 inches apart, and then thin to 8 inches when the seedlings are about 4 inches tall.

Feeding biweekly with a high-nitrogen fertilizer, such as fish emulsion, will keep the plants growing until the silks are set. Or you could try the Native American technique of burying a fish head with the seed. Give the plants plenty of water, especially as the ears start to swell. Don't worry if you see roots popping above soil level; these are a means of anchoring the tall plants.

If corn borers are a problem, control them with *Bacillus thuringiensis* (Bt). Watch for a grayish fungus called smut. It's a delicacy in some parts of the world, but remove it if you want the unadulterated corn.

HARVESTING Look for fat ears with brown tassels. Test corn for ripeness by piercing a kernel with your fingernail: if the liquid is milky, it's ready. Cook as soon after picking as possible, because the sugars start turning to starch immediately. You'll get only a couple of ears per plant, but you can extend the harvest by planting varieties that mature at different times.

VARIETIES *Early*: **'Trinity'** (60 to 70 days) produces a small plant and ear, but it's sweet and early. *Midseason*: **'Tender Vision'** (75 days) is one of the best-tasting varieties. *Late*: **'Silver Queen'** (85 to 90 days) and **'Silver King'** (80 to 85 days) are large, sugar-enhanced varieties that are worth the wait.

Cucumbers

You'll need only a couple of cucumber vines to keep your family in salads all summer long. In good years, you'll be able to supply the whole neighborhood. Vining types can be allowed to sprawl on the ground or trained up a trellis. Bush varieties can also spread several feet.

GROWING Cucumber seed is usually direct sown. They are heat lovers, so wait until after your last frost date and the soil has warmed to about 60°F. If you want a head start indoors, sow seed 3 or 4 weeks before transplanting, because older seedlings do not transplant well.

In the garden, keep the plants well watered; drought can cause a bitter taste. It can take a while for fruits to begin setting. Cucumbers produce both male and female flowers. The males start blooming first, so don't be dismayed if at first you see a lot of flowers and no fruit. You'll know the female flowers have started when you see small fruits forming where the flowers attach to the stem. Cucumbers, like most vine crops, are also heavy feeders. Start with a rich soil and feed them a balanced fertilizer once a month.

HARVESTING Harvest when the fruits have reached their mature size. If they remain on the plant too long, they will turn yellow and become bitter.

VARIETIES **'Marketmore'** (60 to 65 days) is disease resistant and prolific over a long season. Crunchy, sweet **'Orient Express'** (65 days) is low on seeds and high on yield. **'Lemon'** cukes (65 days) are yellow, but not bitter, and they make great little edible bowls. You don't need to peel **'Armenian'** (50 to 60 days), which thrives in hot, poor conditions.

Currants and Gooseberries

If you are looking for an easy berry for edible landscaping, check out the ribes—currants and gooseberries. Bushes sport attractively lobed leaves and glossy berries, and they grow to 3 or 4 feet tall, so they can be grown in containers. They can also be trained into standards. Ribes species were banned for a while because they are a host for white pine blister rust, but most states now allow their cultivation. Some still prohibit growing black currants (*R. nigrum*), however, so check before planting. Resistant varieties are generally allowed.

GROWING Both currants and gooseberries like to grow where it's cool and moist. They will happily tolerate a partially shaded site. The best planting times are in the fall or very early spring, because they leaf out and flower early. Start with 1- or 2-year-old bushes, spaced 3 to 5 feet apart. Ribes can be trellised for improved sun exposure and air circulation, or they can be grown as shrubs. Mulch well and water regularly. These heavy feeders like rich soil, so add yearly side dressings of compost or manure.

Both currants and gooseberries fruit on 2- and 3-year-old branches, so you'll need to do some pruning to keep them productive. Prune the plants in late winter or early spring while they are dormant. After the first year, prune all but six to eight of the strongest shoots. After the second year, remove all but four to six of the current year's canes and three or four 2-year-old canes. Continue this until all you have are three or four 1- and 2-year-old canes.

HARVESTING Currants and gooseberries ripen over a 2-week span, usually in June. They change color as they ripen, but you can sample a few to determine when they're ready to eat.

VARIETIES *Currants*: **'Rovada'** produces large clusters of sweet red berries. **'Translucent Primus'** is a delicious but hard to find white currant. Stick with resistant **'Consort'** and **'Crusader'** for black currants. *Gooseberries*: **'Poorman'** is large, easy growing, sweet, and productive. **'Hinnonmaki Red'** and **'Hinnonmaki Yellow'** are also nice choices for home gardens.

Eggplant

Eggplants are heat worshippers that need a long growing season; the large, voluptuous purple types can be difficult to mature in colder regions. Luckily, because there are all sizes and shapes of eggplants, you can probably find one that will do well in your garden, whatever your climate.

GROWING Start seeds indoors, 6 to 8 weeks before their transplant date. Plant ¼ inch deep and keep the pots warm, at least 80°F, for best germination. Hold off transplanting until there is no chance of frost and the soil temperature reaches 60°F. Preheating the soil with overlaid black plastic can buy you a couple of extra weeks of growing, or consider planting them in containers, which will hold the heat and can be moved if necessary. Space transplants 18 to 24 inches apart. Stake the plants at transplant time, because even a few fruits can be heavy enough to break a stem. Although eggplants are heavy feeders, too much nitrogen can mean fewer fruits, so use a balanced fertilizer every month or so.

Flea beetles will riddle the leaves with holes but ignore the fruits. This can weaken the plants, and row covers are advised for bad infestations. Remember to remove the row covers as soon as flowers are present, because they require pollination.

HARVESTING Ripe eggplants have firm, glossy skins. Overripe fruits will quickly turn bitter and pithy. Cut the fruits from the plant and beware of the prickly stems.

VARIETIES The gorgeous purple-streaked heirloom **'Violette di Firenze'** (90 to 95 days) produces tender, orb-shaped fruits. **'Diamond'** (65 to 90 days) produces long, sweet fruits. **'Hansel'** (55 days) fruits grow in clusters and can be harvested when they are only 3 or 4 inches long, so they are great for short seasons. **'Gretel'** (55 to 60 days) is the white version, with lots of pale, nonbitter fruits. **'Japanese White Egg'** (65 days) resembles eggs and will remind you how this vegetable got its name. The small, white fruits are mildly flavored with few seeds.

Fava Beans

A much-anticipated spring treat, fava beans, or broad beans, have a delicious flavor and grow easily in the cool of spring. The plants are quite ornamental, with lush, green leaves; square stems; and pretty white flowers, each with a black dot.

GROWING Direct sow in the spring, as soon as the soil can be worked. If the soil is still cool, use inoculants to help the plant roots fix nitrogen. Plants need a prolonged period of cool weather, and gardeners in zone 7 will have better luck with a fall crop. Soaking seeds for an hour or two before planting speeds germination but is not required. Plant seeds 2 inches deep, spaced 4 to 6 inches apart. Thin seedlings to 8 to 12 inches to improve air circulation. Once the plants start setting beans, some gardeners pinch or cut off the top 3 or 4 inches of the stem to encourage an earlier harvest and to thwart disease and insect pests by opening the plants to air. Taller varieties may need staking.

HARVESTING When the pods look plump, but not bumpy, the beans are ready to harvest. They tend to hang on, so cut rather than pull them from the plants. Getting them out of their shells requires some work: Remove them from the pod and then pop the seeds free of their outer membrane. If the membranes are difficult to remove, blanch the beans and try again. The tender, leafy green tips are also edible.

VARIETIES **'Broad Windsor'** (80 days) has large, tasty beans and is very reliable. **'Negreta'** (70 days) is an extra early Italian variety. **'Statissa'** (70 days) produces lots of pods on compact plants.

Fennel

Bulbing fennel combines a sweet, warm licorice flavor with the juicy crunch of celery. Bulbs are wonderful when eaten fresh as a snack or in salads, and they're delicious when sliced and caramelized by grilling or broiling. Fennel has become associated with fall, but it is easy and quick to grow throughout the season.

GROWING Start seed indoors 4 to 6 weeks before your last frost date, and transplant outside after that date. I recommend using peat or paper pots, because fennel's taproot does not like to be disturbed. You can succession seed until midsummer. Gardeners in zones 6 and higher can plant another crop in late summer for a fall harvest. Space plants about 12 inches apart in the garden. Fennel likes a sunny spot and rich, well-draining soil. You can also grow fennel in containers. Provide plenty of water, and side dress with compost monthly to keep it growing. Mulching around the plants will keep the roots cool.

HARVESTING Start harvesting when the bulbs reach the size of tennis balls. The days to maturity will vary with the weather.

VARIETIES You'll often find seed with no variety name, labeled only as bulbing fennel or finocchio. Nevertheless, a couple of nice varieties to look for are **'Orion'** (80 days), a fairly quick grower with a sweet flavor, and **'Zefa Fino'** (80 days), a reliable bulb with a smooth anise flavor.

Garlic

Can you imagine a kitchen without garlic? I can't. Softneck garlic is the type we see in pretty braids and on the grocery store shelves. The bulbs can last for months. Hardneck garlic has stiff stems and larger but fewer cloves than softneck. It is the garlic of choice for cold-weather gardeners who plant in the fall and harvest the following summer. You might also know of the mild-flavored elephant garlic, which is not really garlic at all, but a bulbing leek.

GROWING Wait to plant garlic until after a light frost, or the bulbs will sprout and waste precious energy—and they need to grow for a long time. Plant them in a sunny spot in soil amended with organic matter. Gardeners in warmer zones can try planting softneck garlic in the spring. Plant individual garlic cloves, not the whole bulbs, in 2-inch-deep holes, with the pointed ends facing up. Space them 3 or 4 inches apart in rows spaced at least 12 inches apart. Elephant garlic needs to be planted deeper, at 4 to 6 inches, and farther apart, at 6 to 8 inches. Water them well and keep them watered until the ground freezes. Once the ground has frozen in colder climates, mulch the bed with straw, leaves, or evergreen boughs.

Give the area a dose of food in the spring as new shoots emerge, and make sure they are receiving regular water throughout the spring. Hardneck garlic will send up a skinny flower stalk, called a scape, that curls around itself. Harvest the scapes just as they start to bend; leaving them on the plant takes energy away from the growing bulb. Toss the scapes in a sauté; they're delicious. Stop watering when the scapes appear so that the underground bulbs can mature.

HARVESTING When the garlic's lower leaves start to turn brown, dig up a couple of bulbs and test them for readiness. If the cloves have all filled out, they are ready to harvest. To avoid damaging the bulb, dig out the garlic rather than pulling it out. Brush off as much soil as possible and allow the bulbs to air dry and cure for 3 or 4 weeks in a well-ventilated, dry spot. You can cut off the roots and hardneck tops after they have dried and brush off any remaining soil.

Store garlic in a cool, dark location. Softnecks can last 6 to 8 months, but hardnecks should be used within 4 months or they will sprout and go soft.

VARIETIES *Hardneck*: **'Georgian Fire'** (8 to 10 months) is a spicy porcelain garlic that produces about six cloves per bulb. **'Purple Italian Easy Peel'** (8 to 9 months) has a warm spice and handles wet weather well. *Softneck*: **'Creole Red'** (3 to 5 months) is a mild variety with lots of flavor.

Grapes

Grapes are native to North America and the most widely grown fruit in the world. Of course, a lot of the fruit is turned into wine rather than eaten fresh. The vines are not difficult to grow, but getting a good bunch of grapes requires effort. What you do with them after that is up to you.

GROWING Grapes need lots of sunshine and warm weather to ripen and sweeten. Avoid planting them where they will be subject to high winds, which can damage shoots and leaves and diminish fruit set and fertilization. Give them a rich, well-draining soil and plenty of water while the grapes are plumping up. Plant the vines—which come as unrooted cuttings, bare root, or in a container—in early spring, as soon as the ground can be worked. Space them about 6 to 8 feet apart. About 2 weeks after you plant, feed them a high-nitrogen fertilizer such as alfalfa meal, manure, or fish emulsion.

Add whatever type of support you intend to use at planting time, such as an arbor or a trellis. Or you can use the four-arm Kniffen system, the conventional home orchard method: Place tall posts on either side of the main cane, 16 to 24 inches apart, and stretch 12.5-gauge high-tensile wires between the posts, one at about 3 feet and the another at about 5 feet.

HARVESTING Some gardeners let their grapevines run amok, because pruning grapes the proper way can seem like a complicated process. But grape plants are vigorous and forgiving, and even if you make a pruning mistake, you'll get a chance to fix it next season. You can find information on pruning techniques via a plethora of online sources, or you can contact your local extension agency or agricultural service for instructions.

Birds and many animals love to sample grapes, and you may need to install netting to save some fruit for yourself. Few insects cause problems, but grapes can often contract mildews and black rot. If you don't want to use organic fungicides such as neem, choose resistant grape varieties.

VARIETIES **'Himrod'** (zones 5 to 8) is a large, sweet, green, seedless table grape. **'Mars'** (zones 5 to 8) is a seedless blue grape with good cold hardiness and some disease resistance. **'Einset Seedless'** (zones 5 to 8) is a hardy red seedless with a faint hint of strawberry flavor. For red wine, **'Frontenac'** (zones 3 to 8) shows some disease resistance.

Herbs

Only a few herbs are winter hardy in the Northeast, and even those can be iffy in the coldest years. But growing herbs as annuals still provides a big payoff with minimal effort, and you can bring some plants indoors out of the cold weather to enjoy year-round. The woody herbs—such as lavender, oregano, sage, and thyme—tend to be perennial in all but the coldest zones.

Basil. Annual, seed or transplant. Basil needs warm days and lots of water. Start pinching and harvesting when plants are 6 inches tall to encourage bushy side shoots. Some favorites are **'Genovese'**, **'Sweet Dani Lemon'**, and **'Siam Queen'**.

Bay. Woody, transplant. Although bay is hardy only to zone 8, the rest of us can grow it in a container outside and bring it inside each winter. The shrubby tree remains small in a pot. Water weekly and mist if the inside air is dry.

Borage. Annual, seed. With the scent and flavor of cucumbers and its vivid blue flowers, borage is a lovely addition to the garden as well as dishes and drinks. Once it goes to flower, it will self-sow.

Chives. Perennial, transplant or seed. Both chives and garlic chives grow in thick clumps of long, grassy leaves. They can be snipped and used to add flavor to a dish. Beware when they start self-sowing, because they spread quickly. You can divide larger clumps to make more plants.

Cilantro. Annual, seed. Cilantro is tricky in the Northeast. It doesn't like any suggestion of warm weather, and if you can give it moist, partial shade, it will stand a better chance of survival. Otherwise, set out transplants early in the spring and in fall.

Dill. Annual, seed. You'll find fern-leaf type dills and some with broader leaves. It grows with abandon and self-sows. If plants bolt, you can add the seeds to homemade breads and other dishes.

Fennel. Perennial, transplant. Grown for its ferny, anise-flavored leaves, fennel is also a pretty ornamental plant. Temperamental perennials, they are sometimes hardy and sometimes ephemeral. The green variety is easier to grow, but bronze fennel makes a nice contrast in the garden. The seeds of both are edible.

Lavender. Perennial, transplant or seed. We may never have billowy hedges of lavender, but we can grow the plants. English lavender (*Lavandula angustifolia*) varieties, such as **'Munstead'** and **'Hidcote'**, are the hardiest, surviving to zone 4 with protection. Don't prune plants until you see new growth in the spring.

Mint. Perennial, transplant. You can never have just a little mint in the garden, because these plants spread quickly and prolifically, especially in moist shade. To avoid this issue, grow your mint in containers. There are plenty of varieties and flavors, including peppermint, spearmint, apple mint, pineapple mint, lemon mint, and even chocolate mint.

Oregano. Perennial, transplant. Greek oregano is the variety sold for cooking, but many other types are flavorful, and you should feel free to experiment. Plants spread and benefit from frequent division to prevent them dying out in the center.

Parsley. Biennial, transplant or seed. Homegrown parsley is so tasty it should be moved to the center of the plate. Both the flat-leaf and the curly types will continue growing throughout the season, especially if you keep harvesting stems. Plants may overwinter, but they quickly bolt to seed the second season.

Rosemary. Woody, transplant. Like bay, rosemary is hardy only to about zone 8. But in a container, it stays compact and is easy to move indoors and out. With a distinctive resinous, evergreen scent and flavor, a little goes a long way. Keep it misted indoors while the heat is on, but watch out for powdery mildew.

Sage. Woody, transplant. Sage tends to be a low, sprawling plant. Prune it fairly hard after flowering to keep it from becoming too woody. Common sage has the strongest flavor and the hardiest disposition, but the golden, purple, and tricolored sages are tasty, too. Pineapple and other fruit sages are related, but they are not woody or terribly hardy. They can be grown as annuals or houseplants.

Salad burnet. Perennial, transplant or seed. This herb is actually a member of the rose family. Its abundant leaves have a nice flavor that's similar to cucumber, but this salad ingredient is available much earlier. Cut back after flowering for a new spurt of growth.

Thyme. Perennial, transplant. Although the plants and leaves never get very large, thyme has a minty, herbal flavor that is versatile in the kitchen. It also comes in several flavored varieties, with lemon thyme being the most popular. Like oregano, thyme needs occasional division to rejuvenate the plants.

Kale

For a beautiful plant that just keeps going throughout summer and well beyond, you can't beat kale. The plants are large, almost shrub size. Young kale can be eaten raw, and large leaves can be cooked just about any way you can imagine, even baked into chips to fool the kids.

GROWING As with all brassicas, kale prefers cool weather. You can start seed indoors 5 or 6 weeks before your last frost date and transplant out after danger of frost. Cover seeds with ½ inch of soil and keep them moist. They germinate very quickly. In the garden, space transplants 16 to 20 inches apart. You can direct sow as soon as the soil can be worked. Although kale plants can keep producing for 2 months or longer, if you are harvesting regularly, succession plant every month or so. Zone 7 gardeners will have better luck with a summer-sown crop that matures into fall.

HARVESTING Harvest a few tender, young leaves for salad, but leave enough for the plants to keep growing. Harvest the outer leaves first.

VARIETIES **'Red Russian'** (60 days) has hearty, purple-tinged leaves and is rarely bothered by insects. The lacinato (or dinosaur) varieties, such as **'Nero Di Toscana'** (60 days), are great for simmering.

Kohlrabi

Kohlrabi's bizarre appearance could have every gardener asking, what is it? And where can I get it? As it hovers over the soil like a space alien having a bad hair day, you might never guess that kohlrabi has sweet, crunchy flesh with a fruity, cabbage-like taste.

GROWING Sow seed indoors, 4 to 6 weeks before your last frost. Cover with ¼ inch of soil and keep moist. Transplant after danger of frost, spacing the plants 4 to 8 inches apart. Succession plant until midspring, and start a fall crop in midsummer. Kohlrabi likes a rich soil and sun to partial shade. Eventually, the leaves will self-shade the plants. Keep the plants well watered to maintain their juicy tenderness.

HARVESTING Harvest spring kohlrabi when it is about 2 or 3 inches in diameter, before the heat makes it woody. Plants growing into fall can be allowed to get larger, but don't let them start to crack, which opens the plants to disease and insect infestation. Kohlrabi can handle light frost, and, like most of the brassicas, it gets even sweeter for the exposure. The whole plant is edible and is great fresh or cooked in any number of ways; you can even cook and mash the bulb.

VARIETIES Standard varieties **'Early Purple Vienna'** and **'Early White Vienna'** (both 55 days) are reliable, creamy, and mild. **'Kossak'** (65 to 80 days) can reach up to 8 inches in diameter and still be sweet. The outside may get woody, but inside it's fine. **'Winner'** (45 days) is a pale green variety with an intense fruitiness.

Leeks

As beautiful as any ornamental grass, the blue-green leaves of leeks add an architectural element to the vegetable garden. Good thing, because leeks require a long growing season. From starting them in the bleakness of winter to harvesting 6 months later, leeks can really build your anticipation. They seldom disappoint.

GROWING Start seeds indoors 8 to 12 weeks before your last spring frost date. In zone 7, start leek seeds in midsummer to transplant in fall for a spring harvest. Leek seedlings look like thin blades of grass. When they are about 3 inches tall, thin them to one or two plants per pot and allow them to grow and bulk up indoors. When they grow to the diameter of a pencil, they are ready for the trip outside in well-amended, rich soil. If you are planting a row of leeks, it is easiest to dig a 6-inch-deep trench, place the leeks in the trench spaced 4 inches apart, and backfill with only about 2 inches of soil. To get the tender white leaves, you'll need to blanch leeks. Start blanching 4 to 6 weeks after transplanting by adding more soil to the trench or by mounding soil or straw at their base. Do this two more times as they grow taller. After leeks reach ½ inch in diameter, thin the row so that plants are spaced 6 to 8 inches apart; you can eat the thinnings. You can leave leeks in the ground through fall and winter, but mulch them after the first hard frost to prevent heaving.

HARVESTING You can start harvesting your leeks when they reach 1 or 2 inches in diameter. Use a fork to loosen the soil and lift the plants out, roots and all, to avoid accidently snapping off the tender, blanched base of the leek. Harvest only what you plan to use immediately. Leave the rest of your leeks in the ground, to harvest as needed.

VARIETIES **'Blue Solaise'** (175 to 200 days) is ideal for short seasons. In cold weather, the leaves will deepen in color but remain delicious. **'Carentan'** (155 to 215 days) has a delicate herbal flavor but a rugged hardiness. The mild Scottish heirloom **'Giant Musselburgh'** (135 to 230 days) will do well almost anywhere.

Lettuce and Salad Greens

Many people are surprised the first time they experience the sweetness of fresh-picked lettuce. Of course, many people were surprised the first time their plate of salad showed up with reds and ruffles mixed in with the greens. Most salad greens are quick growers that can be tucked in any available spot in the garden.

GROWING You can get a head start on the season by sowing seed indoors, 4 to 6 weeks before your last frost date, but lettuce is a quick grower, and direct-sown seed will soon catch up with transplants. Start sowing outdoors a couple of weeks before your last frost date. You are growing lettuce for its leaves, so provide a rich planting bed. Barely cover the seeds with soil. They need light to germinate.

Eventually thin the seedlings to 4 inches apart. The key to growing great lettuce is water, water, and more water, especially when the weather warms. Succession plant every 3 or 4 weeks throughout the summer. Spring lettuce will appreciate full sun, but during the summer, plant lettuce in the shade of taller plants.

HARVESTING Loose-leaf varieties will keep producing over a long period if you pick the outer leaves and let the center continue growing. Head lettuce is ready when it feels full and solid.

VARIETIES *Loose-leaf*: Grows in fluffy heads or rosettes; leaves can be harvested individually. **'Lollo Rosso'** (45 to 65 days) is heat tolerant with wavy, green leaves tipped with red. **'Deer Tongue'** (45 to 50 days) is sweet, with tender, strappy leaves. **'Black Seeded Simpson'** (30 to 45 days) has been grown forever for its reliable heads of delicate leaves. I grow **'Merlot'** (30 to 55 days) just to look at its stained-glass-red leaves, but it doesn't hurt that it's juicy-sweet, too.

Bibb/butterhead: Leaves have a smooth, buttery texture but few calories and no fat. **'Merveille des Quatre Saisons'** (50 to 55 days) may be a marvel for only three seasons in the garden, but it produces well throughout the summer, when the red color intensifies. **'Tom Thumb'** (40 to 60 days) delivers single-serving heads.

Romaine (cos): Leaves on the loose, upright head have a crunchy rib and a pleasantly bitter flavor. **'Speckled Trout Back'**, or **'Forellenschuss'** (55 days), offers flavorful, apple-green leaves splashed with maroon. **'Rouge d'Hiver'** (55 days) has a crisp green heart surrounded by tall, wine-colored leaves.

Crisphead: Leaves cup over to form a tight, crispy ball. Crisphead lettuces mature slower than other varieties. **'Great Lakes'** (75 days) is easily recognizable as the standard for iceberg, with pale, dense, delightfully crunchy heads. **'Crystal'** (55 to 60 days) is a quick header with glossy green leaves shaded in red. The French heirloom **'Reine des Glaces'** (Ice Queen) (65 to 70 days) is a slow bolter that stays crisp in heat.

Other salad greens: Don't let your salads stop with lettuce. Start the year with mache, or corn salad (40 to 60 days), with its nutty flavor and buttery leaves. Sow in fall or late winter and it will be the first thing up and growing in spring. The red mustard **'Osaka Purple'** (20 to 45 days) is tangy and tender when picked young. Plant broadleaf cress (20 to 30 days) in early spring for a snappy mustard flavor. Or try a microgreen mix (5 to 15 days) of beets, kale, broccoli, and mizuna to punch up a salad any time of year. You can grow these split-second greens indoors or out.

Melons

What would summer be without the sweet, cooling indulgence of melons? We may not love the hot, hazy days of summer, but melons do. The more sunshine they get, the sweeter they become.

GROWING Melons require warm, 65 to 75°F, soil, which is usually the case when the peonies bloom. You can speed things along by covering the soil with black plastic to warm it up. I like to plant melons alongside my driveway to take advantage of the heat it reflects. You can even grow melons in large plastic containers that warm in the sun. Melon seeds germinate quickly. If you want to start seeds indoors, sow 3 or 4 weeks before you plan to transplant. This is a good idea for some of the longer season melons, such as watermelon, but use paper or peat pots to avoid disturbing their sensitive roots when you transplant.

Melons are hungry plants. Work in several inches of compost or rotted manure before planting. To direct sow, make a small mound, or hill, and plant five or six seeds per hill, 1 inch deep. Thin to three plants per hill when the seedlings start to leaf out.

I like to grow young melons under row covers. This serves two purposes: it keeps the insects at bay and keeps the soil temperature warm. I remove the covers when the plants start to bloom so the flowers can be pollinated. Squash bugs and cucumber beetles love melons, and the row covers will foil them early in the season, but be on guard for egg masses on the underside of leaves or clustered inside flowers; hand pick the adults. Powdery mildew also attacks melon vines and can stress growing plants. Spray a mix of 3 parts milk to 6 parts water every 2 weeks, from mid-July through harvest, to prevent it.

HARVESTING Muskmelons (cantaloupes) are the easiest melons to harvest. When ripe, they slip, or separate from their stems. Honeydews and watermelons are trickier. The rind color will change as they ripen, and honeydews will start to smell sweeter. Judging watermelons is something of an art, but one indication of readiness is when the curling tendril on the vine nearest the watermelon starts to turn brown and dry.

VARIETIES *Muskmelon*: The heirloom **'Hale's Best'** (80 days) is aromatic and sweet and tolerant of cool weather. **'Minnesota Midget'** (60 days) has both compact vines and small, softball-sized fruits, but it's big on flavor. **'Sweet Granite'** (70 days) was developed specifically for home gardeners in cooler, coastal, and mountain climates. *Honeydew*: **'Earlidew'** (80 days) is true to its name, with softball-sized, pale green melons. The large, scrambling vines of sweet **'Orange Flesh'** (85 days) endure drought. **'Venus Hybrid'** (88 days) is very sweet and another great choice for short seasons and impatient gardeners. *Watermelon*: **'Chris Cross'** (85 to 90 days) is an Iowa heirloom that delivers lots of 15-pound (or larger) melons. **'Raspa'** (80 to 85 days) bears blocky, 20-pound fruits in an impressively short amount of time. The name **'Orange Tendersweet'** (90 days) sums up this orange-fleshed, 30-pound watermelon.

Okra

One look at okra's captivating blooms and you'll know it's in the hibiscus family. But okra is not just a pretty face; its furry pods have a tender crunch and a nutty, earthy flavor. It's actually a tropical plant, but we can grow it as an annual in the Northeast. Each plant will reward you with copious amounts of tasty pods.

GROWING Direct sow or start seeds indoors. Direct sow 2 to 4 weeks after your last frost date. If starting indoors, plant seeds 2 or 3 weeks before transplanting, and don't even think about transplanting them outdoors until there is absolutely no chance of freezing temperatures. Okra does not transplant well, so use paper or peat pots and plant the entire pot. The large okra seeds have a hard shell, and soaking them overnight will speed germination. Plant them ½ to 1 inch deep and space transplants 10 or 12 inches apart in the garden.

Okra is a heat lover and doesn't really kick into gear until midsummer. It's a heavy producer when it lives in fertile, well-drained soil and you side dress plants with compost when the first pods develop. Keep plants well watered. Few pests bother with okra.

HARVESTING When the pods start coming, you will be harvesting every day as they reach 2 to 4 inches, when they are their most tender and flavorful. Pods grow so quickly you can almost see them elongating. Stay on top of the harvest, because they get tough and woody if they get too large.

VARIETIES **'Red Burgundy'** (55 to 60 days) has stunning foliage and flowers, and its pods are tender even when oversized. **'Clemson Spineless'** (55 to 65 days) is a classic, with tender green pods and no spines. The overachiever **'Cow Horn'** (55 to 65 days) churns out extra-long, 6- to 8-inch pods. Louisiana heirloom **'Stewart's Zeebest'** (65 to 70 days) has unusual round, rather than ribbed, pods that stay tender longer.

Onions

Onions are a kitchen staple; I can't imagine life without them. The beauty of growing your own is the variety that is available. The grocery store offers red or yellow, but you can grow torpedo, walking, and potato onions in your garden. Can we ever have too many onions in the garden?

GROWING If you start from seed, you'll find the biggest selection. Or you can purchase sets of mini onion bulbs or transplants that look like blades of grass. Transplants usually result in larger onions than bulb sets. To start from seed, get going indoors early, 10 to 12 weeks before your last frost date. Sow seeds ¼ inch deep and keep them moist. Give them a boost with some kelp fertilizer or fish emulsion, and when they reach 3 or 4 inches tall, clip their leaves by a third to force energy to the roots. If you prefer bunching onions, you can succession sow them in the garden every 4 to 6 weeks. Sow ½ inch deep and 1 inch apart.

Onions need a loose, well-drained, fertile soil; at least 6 hours of sun; and lots of water. They are shallow rooted and don't like to compete with weeds; otherwise, they are not very demanding. Most pests are repelled by the scent of onions, and I interplant them with other vegetables just to confuse the intruders. The exception is onion root maggots, which burrow into the bulbs; rotating where you plant them each year will help avoid this pest.

HARVESTING Onion greens topple over when the bulbs are ready to harvest; at this point, dig a few to test. When they're ready, dig them up; don't pull them out, or you risk tearing off their leaves, without which the onions won't cure well enough to store. Shake off the excess soil and let the greens remain while you cure them for 2 weeks in a warm, well-ventilated spot, out of direct sunlight. When they are fully dry, remove any remaining soil and cut the tops or braid them together. Store them in a dry, well-ventilated area.

VARIETIES For long-storing onions, try **'Copra'** or **'Cortland'** (both 105 days), mild all-purpose onions. **'Long Red Florence'** (100 to 120 days), an Italian heirloom, has an elongated bottle shape and a mellow onion flavor. **'Evergreen'** (60 to 120 days) bunching onions are about the easiest onion you will ever grow.

Parsnips

Underappreciated and underused, sweet and hardy parsnips spend the summer forgotten underground. Come winter, you'll be glad you planted them. Roasted, boiled, or mashed, these homey vegetables go with just about anything.

GROWING Think of parsnips as white carrots: they like a loose, stone-free, well-draining soil and plenty of moisture. Direct sow any time after frost, but waiting until early summer lets the plants mature in cool weather, which sweetens them. Sow seeds ½ inch deep and ½ inch apart, and cover lightly with soil. Keep them watered, because the seedlings won't be able to break through the surface if the soil crusts over. Thin to 3 to 6 inches apart. Like carrots, parsnips are also a favorite of the carrot rust fly.

HARVESTING You can start harvesting in October, but parsnips will be sweeter if you wait until after a frost. You can leave them in the ground through winter if you mulch with a thick layer of straw to keep the ground around them from freezing. Dig them up by loosening the soil with a fork and then pulling the parsnips out.

VARIETIES For huge, tasty roots, **'All American'** (95 to 145 days) is a top contender and virtually pest-free. Stocky **'Guernsey'** (95 to 120 days), or **'Half Long Guernsey'**, is great for roasting. **'Harris Model'** (100 to 120 days) is slow growing but very sweet, and it overwinters well.

Peas

Fresh garden peas are like potato chips: if you pop open a pod to see if they're ready, you won't be able to stop eating them. Edible snow pea and sugar snap pods are winning the popularity war in the garden, but shelling peas make the best snack. If you have room to grow several rows of them, your homegrown frozen peas will taste that much sweeter next winter.

GROWING There is a tradition of planting peas on St. Patrick's Day (March 17), but don't try it if the ground is still covered in snow. Peas are very tolerant of cold weather, however, and you should get seeds in the ground as soon as it can be worked. If your soil is still quite wet, use inoculants or pre-sprout your seeds to keep them from rotting. Peas need to be up and growing by the time the weather warms. If they are in flower and producing, they'll continue on for a couple of months if they are regularly watered. It's next to impossible to get them started in heat, but if you can find a cool, shady spot in midsummer, you might be able to get a fall crop going.

Direct sow seeds 1 inch deep, spaced 2 or 3 inches apart. Growing them in wide rows of 6 to 10 inches will make low-growing peas self-supporting, but a heavy rain is all it takes to knock them to the ground. Add a trellis or a twiggy branch in front of the plants for them to flop on; taller varieties definitely need to be trellised. Peas are legumes and can store nitrogen in their roots, but rich soil will get them off to a good start. You won't need to feed them, but don't forget to water.

The worst pea pest is the aphid. If the vines get too dense, you may not even notice that aphids are there. When you do, you should be able to blast them off with the hose, but it can take a few tries to get rid of them.

HARVESTING Shelling and sugar snap peas are ready when the pods swell and feel full. Snow peas can be harvested when the pod reaches full size, usually 2 or 3 inches, and before the peas inside swell. Keep picking; peas will get starchy as they get old, and the more you pick, the more the vines will produce. You also get a bonus crop with peas: pea tendrils, the tender vine tips, have great flavor and are delicious tossed into salads and stir-fries.

VARIETIES Old shelling favorite **'Lincoln'** (60 to 70 days) grows on short vines, producing six to nine peas per pod. It's tolerant of downy mildew and Fusarium wilt. **'Tall Telephone'** (75 to 85 days), also known as **'Alderman'**, needs a tall, sturdy trellis to support vines that easily reach 6 feet and are covered with tender pods with eight to ten peas. Snow pea **'Golden Sweet'** (60 days) is a beautiful plant with purple flowers, and the buttery yellow pods have a zesty kick. Dependable **'Cascadia'** (58 to 65 days) sugar snap peas grow on short vines.

Peppers

Peppers bide their time until the true heat of summer sets in. Unlike hot peppers, sweet peppers are a little lazy about setting flowers and getting a move on, but they make up for this by being large instead of hot. Peppers can be a challenge in colder New York climates, where the smaller varieties may have to suffice for pepper-loving gardeners. Hot or sweet, thin or stocky, there is a pepper for every garden.

GROWING Sow indoors 8 to 12 weeks before your last frost date. (I said peppers were slow.) The seeds need heat to germinate, so put them in a warm room, on top of the refrigerator, or on a heating mat. Heat will make the soil dry out quickly, so remember to water often. Harden off your seedlings gradually, because the shock of sudden cold can stunt them. Transplant them after temperatures remain reliably above 50°F.

Peppers are not fussy about soil, but the usual rich, well-draining soil is best. Plant them about 1 inch deeper than they were set in their pots; they can send out small roots at the base of their stems. Space plants 14 to 18 inches apart, although I find my peppers do best when I overcrowd and almost abuse them. They are defiant. I think it's wise to stake pepper plants; not all varieties need it, but when they are full of heavy ripening fruits, they can flop over and snap a branch. To be safe, stake them at planting time. Keep the plants well watered, because drought can cause them to drop their flowers, as will prolonged periods of cool weather. Feed the plants with a balanced fertilizer when the first flowers appear.

Peppers, especially the hot ones, aren't bothered by too many pests. Cutworms may try to attack young seedlings and can be thwarted with a collar placed around the base of the plant. Blossom end rot and sun scald are cultural problems that can be avoided with regular watering and leafy growth.

HARVESTING Harvest peppers when they reach the size and color you prefer. (If you like your peppers green, don't apologize.) The more you pick, the more you will get. Cut them off, rather than pulling; a good tug can take half the plant with the pepper.

VARIETIES *Sweet*: Frying pepper **'Banana Supreme'** (68 days) gets high marks for both flavor and performance. **'Giant Marconi'** (70 to 90 days) is an exceptionally sweet bell pepper. **'Golden Bell'** (68 to 70 days) produces well into fall. *Hot*: **'Hot Lemon'** (70 days) is laden with dozens of crisp, medium-hot fruits (5000 to 30,000 Scoville units). The slender, 6-inch fruits of **'Long Cayenne'** (60 to 85 days) dry beautifully (20,000 to 40,000 Scoville units). You'll like the spicy flavor of **'Fatalii'** (90 days), if you can survive the heat (125,000 to 35,000 Scoville units).

Potatoes

There is something very satisfying about digging up potatoes—maybe it's the relief of finding out that they actually grew. Potatoes can take up a lot of space, but more and more gardeners are growing them in barrels or other large containers. I think this is even easier than growing them in the ground. You can use a barrel or a clean plastic garbage can with a few holes drilled at the bottom and around the sides. Add some good soil, and toss in your seed potatoes. Cover them with a few inches of soil, and as the plants grow, keep topping them with soil. At the end of the season, you simply tip over the barrel and dig in.

GROWING We don't grow potatoes from seed; we start with chits, which are potato pieces that include two or three eyes. Unless you've been saving your own disease-free potatoes, you should always plant certified disease-free seed potatoes. Late blight can overwinter in potatoes, and the East Coast has had enough problems with that disease. After you cut up the potato, let the cut pieces sit out overnight to dry and callus.

For most of us, midspring is the time to plant potatoes, and most varieties take 3 or 4 months to develop. You can extend your season by planting early-, mid-, and late-season varieties. As with all long-season crops, starting with soil rich in organic matter and watering regularly are the keys to success. Unstressed, happy plants are less prone to problems.

In addition to the container method, you can try two different potato planting techniques. The traditional method is to dig a shallow trench about 6 to 8 inches deep—allow 2 or 3 feet between trenches if you're planting several. Plant the chits about 10 to 12 inches apart in the trench, eyes facing up, and cover with soil. As the plants grow, continue filling in the trench, burying part of the stems as you go. The nice thing about this method is that the soil does not have time to compact, so at harvest time digging is easier. Keep hilling soil on the plants if potatoes begin popping above the surface. Exposure to sunlight will turn them green and make them mildly toxic with solanine. Stop hilling when the plants flower. You can also go for the lazy gardener method. Lay the potato chits right on top of the soil and dump a shallow layer of mulch on top of them. Then keep layering mulch as the plants grow.

Along with the incurable late blight, potatoes can be pestered by Colorado potato beetles, which will feast on the leaves. Scout for their orange egg masses early in the season to reduce the population. Scab, a disease that causes corky marks and sunken holes on your spuds, can be controlled by lowering the soil pH to 5.0 to 6.0. A lot of potato problems can be avoided by not planting them in a spot where tomatoes or peppers (both in the nightshade family) grew the prior season.

HARVESTING Treat yourself to a few little potatoes after the plants reach about a foot tall. Poke around near the base of the plant and you should find some small nuggets to enjoy. Once the tops of the plants die back, the potatoes are ready.

VARIETIES **'Irish Cobbler'** (90 to 95 days) is an early-season favorite. Yellow-fleshed **'German Butterball'** (90 to 120 days) is creamy and comforting. **'Kennebec'** (70 to 90 days) and **'Katahdin'** (90 to 100 days) store well over the winter. The best blue potato is **'Peruvian Purple'** (100 to 110 days), which keeps its color when cooked. The long and slender **'French Fingerling'** (80 to 100 days) is tender enough not to need peeling.

Radishes

Radishes have been relegated to the salad bar for too long. These spicy red orbs may seem a bit pedestrian, but there is more to radishes than a little color on the salad plate. Hardy winter radishes can fire up stews, add a little jolt to broiled root vegetables, and do a mean cracker imitation. Try these and you will gain new respect for the way those humble little globes wake up a salad.

GROWING Radishes are direct sown in the garden. They can be started very early in the spring as soon as the ground is relatively dry, again toward the end of summer for a fall harvest, and even in late fall with the protection of a cold frame. Radishes can be squeezed in between other plants and are good at loosening and cultivating soil for slow-to-sprout vegetables such as carrots. Spring radishes need to grow quickly. Give them a relatively rich, well-draining soil. Sow seed ¼ inch deep and about 2 inches apart. Thin to 3 or 4 inches when they are 1-inch tall and toss the thinnings in a salad. Because they grow so quickly, plants don't need fertilizing, but make sure they have plenty of water, and don't let them remain dry for several days running. Succession plant every 10 to 14 days for a continual harvest, until the days warm up to the 80s°F.

HARVESTING Fast-maturing spring radishes are ready to eat in 3 or 4 weeks, when they begin to poke their shoulders above the soil. Gently feel around the top of the radish bulb to judge its size. You can pull out the small, round radishes, but it's a good idea to loosen the soil around the long, thin varieties. Don't let mature radishes remain in the ground or they will get pithy or woody, depending on the weather.

Many of the Asian or winter radishes grow slowly. If you plant them in the spring, they won't always have time to mature before bolting. These do best when they're sown in midsummer to late summer. You can harvest them from the fall into winter or let them overwinter in the garden for a spring harvest.

VARIETIES For early spring, the classic, red **'Cherry Belle'** (22 days) and the multihued **'Easter Egg'** (25 days) will get the season off to a spicy start. **'French Breakfast'** (23 days) is a bit more tolerant of heat. Despite its bracing, cool look, the long, white **'Icicle'** (25 days) can also handle some heat. If you want to harvest and eat radishes all summer, try **'Rat's Tail'** radishes (45 to 50 days) for their edible seed pods and great flavor, without the worry of bolting. **'Daikon'** (60 days) is probably the most familiar Asian radish. It has a surprisingly mild flavor and goes very well with beer. Slice up a **'Red Meat'** (45 to 60 days), also known as **'Watermelon'**, radish to use as a cracker for dips and spreads. To overwinter, try the pungent **'Round Black Spanish'** or **'Long Black Spanish'** (both 55 days). The black skin looks woody, but it is tender enough to eat without peeling.

Raspberries & Blackberries

Ripe raspberries and blackberries look like little jewels. The brambles, as they're called, are big, unruly shrubs that are usually pruned and trained for maximum fruiting. Luckily, they tend to ripen over a stretch of time, so you can snack to your heart's content and still have plenty for baking and jams.

GROWING These are hardy plants, and once they take root, you'll find it difficult to get rid of them. If you want only one or two plants, you'll find some decent varieties at nurseries. If you are greedy and want an entire hedge of berries, purchase bare-root plants at a fraction of the cost. Give them a good start with well-amended, well-draining soil. They need full sun to develop the berry's sugars. Space plants or canes 2 or 3 feet apart and cut canes back to 6 inches at planting. Don't worry; more canes will sprout.

There are two basic types of raspberries: summer-bearing, which bear on second year growth, and everbearing, which fruit on new growth in the fall and then on the same canes the following summer. After fruiting, the canes of both types decline and should be cut back at soil level. Which brings us to pruning. For summer-bearing varieties, you want new growth for next year's crop and 2-year-old canes for this year's crop. You can prune out the old canes after fruiting, but you'll still need to do a little maintenance pruning. Thin out any new wood that is thinner than the diameter of a pencil and any canes that cross other canes or that sprout up far from the base of the plant. Everbearing varieties can be pruned the same way after the second fruiting in the summer. If this cyclical pruning gives you a headache, you can cut all the canes to the ground after the fall fruiting and have just one crop each year on new wood.

The same laissez-faire attitude can be applied to trellising. The conventional way to train berries is to run two horizontal wires on either side of the plant and secure the canes to one side or the other, in a kind of running V shape. This keeps the canes from flopping over and opens them to sun and air. But the brambles will fruit even if you let them go au natural.

HARVESTING Ripe berries will change to their mature color and soften, and you can start to smell their sugars. Try a few to test, but the best sign of readiness might be when the birds start hanging out nearby.

VARIETIES *Summer-bearing raspberries*: **'Jewel'** (July) is a black raspberry with glossy, firm, very sweet fruit. **'Reveille'** (July) was developed in Maryland and has great cold hardiness. Best for fresh eating, the soft, juicy fruit ripens in early summer. New Yorker **'Taylor'** (August) has good disease resistance and large, late-ripening, firm berries. *Everbearing raspberries*: Another New Yorker, **'Heritage'** (July and September) is high yielding but thorny. The fruit is excellent and firm enough to stand up to freezing. **'Autumn Bliss'** (August and September) delivers lots of large, full-flavored berries. **'Fall Gold'** (July and August) has luscious, soft, yellow fruits blushed with pink. *Blackberries*: **'Chester'** (August) is a thornless favorite. It is very hardy with some disease resistance. The fruit is on the tart side but delicious. An earlier thornless fruiter, **'Triple Crown'** (July through August) is a sweeter blackberry.

Rhubarb

There was a time when rhubarb was the first sweet treat of the year. After a winter of root crops, the first stalks of rhubarb meant dessert. It's an odd thought that a leafy vegetable can do such a good job of impersonating a juicy fruit, but rhubarb pulls it off.

GROWING Rhubarb is one of a handful of perennial vegetables. When something takes up long-term residence in your garden, you want to make it comfortable. Start with a sunny location and rich soil deeply amended with organic matter. Rhubarb has a short season and needs to grow fast. It is rarely started from seed, because seed does not always grow true, and it can take 2 years to reach harvestable size. Instead, we plant crowns, dense chunks of bare-root divisions. Space the crowns 3 or 4 feet apart in rows about 3 feet apart, with the crown 2 inches below the soil surface.

Keep the plants well watered and allow them to grow and establish themselves the first season. A layer of mulch around the plants will help the soil retain moisture. Remove any flower stalks and continue watering, even if the plants go dormant. Feed in early spring with a balanced fertilizer. You'll need to divide the plants every 2 to 4 years. Crowded plants produce thin stalks. You can slice out the older center of the plant and leave the newer outer edges in place, or dig out the whole root mass and divide the crown into 2-inch pieces, cut between buds.

HARVESTING After transplanting, wait 2 years before harvesting the stalks. The following spring, begin harvesting when the plant has fully leafed out. Grasp the stem at the base and pull down with a twist. Never harvest more than half the leaves of any plant, or it might not have the energy to regrow.

As the weather warms, growth slows down and your rhubarb may go dormant. Stop harvesting when you see little new growth, which is usually around the beginning of June, and let the plants reestablish. Note: Eat only the stalks and never the leaves. The leaves contain toxic levels of oxalic acid. Also do not eat stalks that have been frozen and have fallen over, because the oxalic acid can move from the leaves into the stalks.

VARIETIES A good choice, especially for northern gardeners, is **'Valentine'**, with thick, red stalks that are very sweet and retain their color when cooked. It is also very slow to bolt. The classic rhubarb is **'Victoria'**, which grows with abandon. Two others I like are **'Canada Red'**, with thin and tender stalks, and **'Giant Cherry'**, which does well even in areas with milder winters.

Shallots

If you're like me, you would never think of not planting garlic in the fall, but you don't hesitate to buy expensive shallots at the market. Shallots, the sophisticated onions, are every bit as easy to grow as garlic, and when you have your own stash, you'll find all kinds of uses for them. Try the old chef's trick and sauté them in butter before adding them to your favorite recipe.

GROWING You can grow shallots from seed, but because they rarely set seed, it can be hard to find. Instead, shallot sets, or bulbs, are available in the fall. They need rich, loose soil with a slightly acidic pH. In cooler climates, plant them by mid-October. You could also plant them in the spring, but they will yield smaller bulbs. Gardeners in milder climates can plant in late fall. Plant individual bulbs just below the soil surface, 4 to 6 inches apart. Keep the bulbs well watered until the ground freezes. Do not mulch, because shallots are susceptible to many types of rot and the soil needs to dry between waterings. Side dress with organic matter in early spring. Cut back any flower stalks to focus the plant's energy on the bulb. You can also cut the leaves back by a third for the same reason.

HARVESTING Cut a few of the greens to use like green onions. The bulbs are ready when their tops start to yellow and flop over. Fall-planted shallots will be ripe in early summer; spring-planted shallots will mature later in the summer. Cure shallots for storage by shaking off the excess soil and allowing them to sit in a dry, shaded spot for 2 to 4 weeks. Shallots should store well for up to 8 months if you separate the bulbs and keep them cool and dry. Don't forget to save the largest bulbs to replant.

VARIETIES The French gray (90 to 100 days) is the gourmet's go-to shallot; some consider it the only true shallot. **'Red Sun'** (80 days) is another French shallot, with glowing red skin and spicy, blushed interior rings. **'Ambition'** (100 days) is a good choice for northern gardeners. It's a large French shallot with copper skin and white flesh, and it stores extremely well. **'Prisma'** (100 days) is red inside and out, with a stronger onion flavor than most.

Spinach

Cool, wet springs are perfect for growing a lush patch of dark, leafy spinach. Packed with nutrients and flavor, spinach has become a darling of the salad bar and side dish. You can find it in a variety of shapes, sizes, and textures. Savoy type spinach has thick, crinkly leaves. Baby spinach isn't necessarily the immature leaves; several tender varieties simply never get very big.

GROWING Direct sow spinach in early spring as soon as the soil can be worked. A quick grower and a heavy feeder, spinach needs fertile soil and a shot of high-nitrogen fertilizer, such as fish emulsion or soy meal, when you start harvesting. It is also particular about soil pH and will not thrive in acidic soil, preferring a pH of 6.5 to 7.5.

Sow seed thinly, ½ inch deep and about 1 inch apart, in rows or blocks. Cover lightly with soil and keep moist. Thin plants to about 3 inches and enjoy the thinnings in a salad. Succession plant every 7 to 14 days for a continual harvest.

As the weather warms, spinach slows down and bolts to seed. You can try growing it in a shady spot with plenty of water, but summer spinach often tastes bitter. But keep your seeds handy, because spinach can be sown again in late summer to grow sweet and tender in the cool, short days of fall. Spinach can handle a little frost, and row covers will keep it going longer. Gardeners in zone 7 will be harvesting well into winter. In colder zones, mulch any remaining plants when the ground freezes, and your spinach will resume growing the following spring.

HARVESTING Harvest individual outside leaves in the cut-and-come-again style to get several weeks of spinach. Or slice off an entire mature plant; if you leave about 1 inch of the crown intact, the plant will send up new growth for a second, smaller harvest.

VARIETIES **'Nordic IV'** (40 to 45 days) is a slow-to-bolt choice for spring, with smooth, tender leaves. **'Bordeaux'** (25 to 30 days) has beautiful, tender leaves with red veins, but it tends to bolt quickly. Semi-savoy **'Tyee'** (39 to 45 days) is one of the longest lasting spring spinaches. For fall planting, **'Giant Winter'** (45 days) has semi-savoyed leaves that overwinter well. **'Bloomsdale Long Standing'** (40 to 48 days), another good fall/winter spinach, has succulent savoyed leaves and germinates well in cold soil.

Strawberries

So many fruits and vegetables are seasonal markers. Strawberries announce that June has arrived, although many varieties will bear throughout the summer. Still, those first plump berries of June seem to taste the sweetest. Resist the pithy strawberries sold in winter, no matter how rosy they look, and hold out for the real thing.

GROWING June-bearing strawberries produce one large crop in—you guessed it—June. These berries are great if you want to preserve your harvest. Everbearing strawberries produce two or three harvests during the course of the summer but are not as highly prized as day-neutral berries, which fruit sporadically throughout the growing season. Spring-bearing alpine strawberries produce tiny, sweet berries that make a carefree groundcover.

Prepare your strawberry bed ahead of time by working in lots of organic matter, and adjust it to a slightly acidic pH level of around 6.0. You may find potted strawberry plants at the nursery, but they are often grown from bare-root stock. To plant a bare-root strawberry, dig a hole large enough to accommodate the spreading roots. Hill the soil in the center of the hole and place the plant on the hill, with its crown slightly above soil level so it won't rot. Spread the roots around and down the hill and refill the hole with soil to halfway up the crown.

Space the plants 12 to 15 inches apart, and keep them well watered. Mulching with straw will help conserve moisture and keep the berries off the ground. Now comes the hard part. For the first year, pinch off all the flowers on June-bearing strawberries and all the flowers that form before July 1 on the other varieties. This will delay your gratification but allow the plants to grow stronger. June strawberries send out lots of side runners. Train them outward and press them gently into the soil, where they will root and become new plants. Feed your strawberry bed each spring with a balanced organic fertilizer.

HARVESTING The sweet scent will tell you the berries are ripening. Wait until the entire berry reaches full color before you pick it, and try to get a small portion of the stem with the berry. Handle them gently, because fresh ripe strawberries are fragile.

VARIETIES *June-bearing*: **'Earliglow'** is one of the earliest and sweetest. **'Northeaster'** has a distinct flavor and good disease resistance. **'Jewel'** is probably the most popular strawberry in the region, with large, sweet, full-flavored fruits. (All varieties are harvested in or around the month of June.) *Day-neutral*: Both reliable varieties, **'Tribute'** has the largest fruits but **'Tristar'** is the sweetest. (All day-neutral varieties are harvested continually from June or July through frost.) *Everbearing*: **'Fort Laramie'** and **'Ozark'** are older varieties with good production. **'Quinault'** fruits in a speedy 4 or 5 weeks. (All everbearing varieties are harvested from June through frost.) *Alpine*: **'Alexandra'** produces lots of dark red berries. **'Mignonette'** starts fruiting its first year, from seed. **'Yellow Wonder'** has aromatic, pale yellow fruits. (All Alpine varieties are harvested from late spring through frost.)

Sweet Potatoes

You may not associate New York with growing sweet potatoes, but why not? They take about 4 months to mature, which means you can even wait until May to plant and harvest in September. You might not get the giant tubers you see in the produce aisle, and mine often grow in funky shapes, but they're still delicious. Sweet potatoes are in the same family as morning glory flowers, and we know how well they grow here.

GROWING Work lots of organic matter into your planting bed and amend it to a soil pH of 5.5 to 6.5. Choose a spot that receives at least 6 hours of direct sunlight to keep the plants warm. Sweet potatoes are grown from small rooted pieces of the tuber called slips. Before you plant, wait until the ground has warmed enough to sit on comfortably. Covering the soil with black plastic for a week or two will help heat it up. Raised beds are a great place to plant sweet potatoes, because the soil warms quickly in the spring, drains well, and is light and fluffy for the expanding tubers. Space the slips 12 to 18 inches apart with 3 or 4 feet between rows. This is a vining crop, so give it room to roam. The vines will become a living mulch, but they can be a bit slow to start growing. Keep the area weeded until they can do their job. Work some compost into the soil and you won't need to add supplemental food. Keep the plants well watered until 3 or 4 weeks before harvest, and then stop watering to prevent the tubers from splitting.

HARVESTING Sweet potato leaves make a nice cooking green and can be cut at any time and used like spinach. The tubers are ready to dig when the leaves start to turn yellow. Tubers grow close to the soil surface and have very tender skins, so be gentle when you dig in. Frost will damage the leaves, but the tubers should be fine.

VARIETIES Moist, orange-fleshed **'Centennial'** (90 to 100 days) grows well and tastes great, but some of the tubers are oddly shaped. **'Beauregard'** (90 to 100 days), a reliable quick grower with disease resistance, should do well in cooler climates. **'Georgia Jet'** (100 days) seems happy to grow anywhere; its meaty, orange flesh bakes well.

Swiss Chard

Standing tall and upright in the garden, with colorful ruffled leaves, Swiss chard can look so perfect you may wonder if it's real. This easy vegetable is grown for its succulent leaves, which taste somewhat like a spinach with a crunchy midrib. Although it does not form a bulb, chard is actually in the beet family, and it's sometimes called silverbeet. Swiss chard loves to grow, and you can get an entire season of harvest from your plants, although I like to start a second crop in midsummer. I guess I'm greedy.

GROWING You can sow chard indoors or out. If you want to get a head start indoors, sow seeds 3 or 4 weeks before your last frost date. Direct sow in the garden about 2 weeks before the last frost date. Choose a site with full sun to partial shade, and amend the soil with organic matter. Chard likes soil on the acidic side, preferably with a pH of 6.0 to 6.4, but it can be forgiving. Like beet seed, chard seed comes in clusters, so some thinning after planting will be necessary. Sow seeds ½ inch deep and 2 to 4 inches apart. Thin the young seedlings when they reach 8 to 10 inches and toss them in a stir-fry or salad. Provide plenty of water and mulch to keep the soil moist and the leaves clean. A midseason side dressing with compost or rotted manure will keep them chugging along. You can start a fall crop by sowing in mid-August. Chard can handle a little light frost.

By far the biggest pest for chard is deer, which see a patch of chard as a personal salad bar. Do not be tempted to grow it in the flower bed, unless you want to feed the deer, rabbits, groundhogs, and other hungry animals instead of your family. Leaf miners can burrow into the leaves; watch for their eggs and destroy them. Slugs also love the leaves and will turn plants from Swiss chard to Swiss cheese.

HARVESTING Harvest the leaves at any size, while they are still glossy. Snap or cut off the leaves at the base and leave the crown intact. Take only two or three leaves from each plant at a time, and more leaves will fill in.

VARIETIES **'Five Color'** (60 days), also known as **'Bright Lights'**, creates a beautiful mix of colors and textures. **'Rhubarb'** (55 to 60 days) defies all kinds of weather. **'Fordhook Giant'** (55 to 58 days) is a great choice for fall growing. **'Perpetual'** (50 to 60 days) tastes a lot like spinach, with smooth, tender leaves.

Tomatillos and Ground Cherries

These two cousins in the *Physalis* genus are very similar in appearance and growing conditions, but they're completely different in flavor and use in the kitchen. Both grow into fruits that are individually wrapped in papery husks. Tomatillos burst open when they are ready to harvest. They have a fruity, slightly tart, tomato flavor and are a mainstay of salsas. They also taste great when grilled or diced into stews. Ground cherries, or cape gooseberries, are sweet little fruits with a pineapple-like flavor. When ripe, they completely fall off their plant and pop out of their husks. Although they make good jams and bake well, I think few ever make it out of the garden uneaten.

GROWING These are easy-growing plants. Give them sun, heat, and water and they are happy. Both tomatillos and ground cherries are very sensitive to cold. You may not be able to find seedlings, but the seeds are easy to start indoors about 4 weeks before your last frost date. Transplant seedlings when the ground is warm enough to sit on. You can also direct sow seed in the garden after danger of frost has passed.

You will need to plant at least two tomatillo plants, because they are self-sterile and need a pollinator to set fruit. Tomatillos get tall and heavy, so staking is recommended. Ground cherries will sprawl across the ground, or you can grow them in containers.

HARVESTING When tomatillos start to plump up, their husks split open. For full sweetness, wait to harvest them until they have fully colored, either green or purple, depending on the variety. The fruits are covered in a sticky film that washes off easily. There is no guessing with ground cherries, because they almost jump off the plant and into your hand. When they drop, they are ripe. Don't eat unripe ground cherry fruits, because they are toxic with solanine, as are the stem, husk, and leaf.

VARIETIES *Tomatillos*: **'Purple'** (75 to 85 days) is a beautiful heirloom with large, sweet, dark purple fruits. Another purple, **'de Milpa'** (65 days), has a pleasant sweet-tart flavor. **'Verde Puebla'** (75 to 80 days) is a classic large green variety. *Ground cherries*: **'Cossack Pineapple'** (65 days) can be eaten fresh, cooked, or dried. **'Aunt Molly's'** (65 to 70 days) is widely available, extremely sweet, and prolific.

Tomatoes

Is there a vegetable gardener who does not grow tomatoes? I think tomatoes are how we judge the growing season's success: if the tomatoes were good, nothing else matters. Ever since heirlooms became mainstream, the number of available tomato varieties has become overwhelming, but there are worse problems to have. You could probably try a new variety every year, and I suspect each would offer something special. It certainly would be a fun experiment. Tomatoes are tropical heat lovers, and they can pose a challenge for gardeners in zone 4 or lower, but do not accept defeat. There is a tomato for your garden, too.

GROWING Seeds offer a wider selection than nursery-grown seedlings. Tomato seeds are quick growers, and you want short, stocky transplants, so don't start them too early—about 6 weeks before you plan to move them out is fine. Sow the seed ½ inch deep. The seeds will germinate faster if the soil is kept at a warm 70 to 80°F. You can either heat the containers from the bottom with plant heating mats or coils or keep them in a heated room. Be sure to move them off the heating mats once they are up and growing, so the soil does not continually dry out. When the seedlings have true leaves, thin each cell or pot to the one strongest seedling by cutting the others at the soil line. To get stocky plants, either use a gentle fan to blow air on the plants or rub your hand gently across the plants for a few seconds every day to strengthen the stems.

Before you put seedlings in the ground, harden off the plants slowly over the course of a week or two. Do not rush transplanting them into the garden: wait until nighttime temperatures are reliably above 50°F. Then plant them in full sun, in a rich soil with a slightly acidic pH.

All tomatoes need some type of heavy-duty staking or caging. Even bamboo poles are not strong enough to support dozens of beefsteak tomatoes. Install your supports at planting time and train or tie the vines as they grow.

HARVESTING Harvest when the tomato has reached full color; it should be slightly soft and very aromatic. We are all faced with the dilemma of what to do with late-season green tomatoes. By early September, your plants have finished setting viable fruits, so remove all the remaining flowers and prune back the top of the plants to force them to focus on the existing tomatoes. If a frost threatens, harvest everything and bring the fruits indoors to finish ripening.

VARIETIES Favorite tomatoes are a personal thing. I favor the purple tomatoes, such as **'Pruden's Purple'** (70 to 80 days), **'Cherokee Purple'** (75 to 85 days), and **'Black Krim'** (70 to 80 days). I like their combination of smoky, sweet, and tart. **'Spitfire'** (68 days) is an early, all-purpose red tomato, good for short seasons. **'Polish'** (85 days) is an exceptionally meaty beefsteak. If you lean toward low acid, fruity tomatoes, try some of the yellow and green varieties: **'Carolina Gold'** (80 days) and **'Green Zebra'** (75 days) are good growers. Small-fruited tomatoes are some of the most prolific: **'Chocolate Cherry'** (70 days) has a smokiness I love. Mild and sweet **'Beam's Yellow Pear'** (70 to 80 days) is a big hit with kids. For sauces, I like **'Opalka'** (75 to 85 days), with its huge yields and meaty goodness, and the mini-Roma **'Juliet'** (60 days), which is as good fresh as it is sauced.

Turnips and Rutabagas

These vegetables get no respect. Even the commercial farmers seem to consider them fillers. For years we have abused them in the kitchen, boiling them into an unpleasant smelling mush, but these versatile vegetables offer more: cook their tangy leaves, roast and caramelize the bulbs, bake them into casseroles and soufflés, or serve them up as French fries. I hope I've tempted you to try growing turnips and rutabagas.

GROWING Turnips and rutabagas like full sun, rich and well-draining soil, and a slightly acidic pH. Because they are bulbs, both need to be direct sown. Turnips are quick growers and can be seeded in early spring or midsummer, about 2 months before your first fall frost. Rutabagas grow more slowly, and a spring sowing will mature in fall. Thin seedlings to 2 to 4 inches apart, and add the thinned plants to a salad. Make sure the plants get at least an inch of water weekly or they will crack, leaving space for disease and insects to enter.

Turnips don't need feeding, but rutabagas like a top dressing of organic matter at midseason. The only serious pest of concern is the root maggot. A row cover will prevent the moth from laying its eggs on the soil.

HARVESTING You can harvest the greens any time, and as long as you don't injure the top of the bulb, the greens will resprout. Turnip bulbs are most tender when small, at 2 or 3 inches in diameter. Fall-planted turnips will sweeten even more if they're overwintered; add a layer of mulch after the ground has frozen. Rutabagas, on the other hand, will not overwinter, but you should wait to harvest them until after the tops have been hit by a frost, which sweetens the bulbs. Mature bulbs are best at 3 to 5 inches in diameter.

VARIETIES *Turnip*: **'Gold Ball'** (45 days) is sweet and tender when bulbs are about 3 inches in diameter and is wonderful raw in salads. **'Shogoin'** (55 to 70 days) is a bright white turnip that is excellent when lightly cooked. *Rutabaga*: Creamy, yellow-fleshed **'American Purple Top'** (90 days) grows huge and stores well. Sweet, white-fleshed **'Gilfeather'** (75 to 100 days) tastes like a turnip.

Winter Squash, Pumpkins, and Gourds

There is something so delightfully comforting about a row of winter squash in your cellar, waiting to be turned into something warm and delicious in the kitchen. Dressed in their fall colors, pumpkins, butternuts, acorns, delicata, and even the inedible gourds dress up the garden, the table, and often the yard. They are large plants, to be sure, but you don't need many to fill your home with color and the wonderful scent of baking pies.

GROWING You can start seed indoors, about 3 or 4 weeks before your last frost date, but direct-sown seed quickly catches up. Direct sow in the garden after all danger of frost. Winter squash need heat to start growing in earnest. Gardeners in zones 3 to 5 can spread black plastic on the soil to warm it quickly. Plants also need a rich, well-draining soil and plenty of sun. Plant seeds much as you'd plant melons and cucumbers, in hills or small mounds. Plant three or four seeds per hill, 1 or 2 inches deep. Space the hills 4 feet apart for bush varieties and 6 feet apart for vines. Thin the hills to the one strongest seedling after they leaf out. Keep plants well watered, especially when the fruits start forming. Winter squash self mulches with its enormous vines.

HARVESTING Winter squash will change to its mature color as it ripens. When your fingernail can no longer dent the rind, the squash is ready to pick. Cut it from the vine, leaving a few inches of stem attached. This is not a handle; keeping the stem attached prevents rotting and disease at the tender end. Let the squash sit and cure in the sun for 2 or 3 weeks, and then bring them indoors and store in a cool, dark spot until you are ready to cook something fabulous during the winter.

VARIETIES *Winter squash*: **'Bush Delicata'** (80 days) is a compact grower with full-sized, creamy fruits. **'Buttercup'** (90 days) is a rambling vine with luscious fruits that store for months. **'Sweet Dumpling'** (90 to 100 days) has an earthy sweet potato flavor and a gorgeous speckled shell. **'Potimarron'** (85 to 95 days) hints at the flavor of chestnuts. *Pumpkins*: **'New England Pie'** (105 days) loves Northeast weather and makes a nice pie or side dish. For sheer beauty, there is **'Rouge Vif D'Etampes'** (115 days), the deep red-orange Cinderella pumpkin. *Gourds*: For fun, try **'Harrow-smith Select'** (95 days), a mix of funky, colorful, bumpy gourds. Or grow some birdhouse or bottle (125 days) gourds; dry them over the winter and create birdhouses with them for the garden next year.

Zucchini and Summer Squash

Anyone can grow zucchini. In fact, zucchini may be able to grow without our assistance. So instead of looking for highly prolific zucchini, which seems redundant, focus on finding the most delicious types. Summer squash should be tender when harvested; although it is true they become sponges for other flavors, they can bring a flavor of their own to a dish.

GROWING Zucchini and summer squash are grown much the same as cucumbers, meaning they pretty much grow themselves. You can start seed indoors, but there is really no need. Wait until the soil has warmed and direct sow in the garden. Use the hill method and plant two or three seeds per hill about 1 or 2 inches deep. Space the hills 2 or 3 feet apart. When the seedlings leaf out, thin to one or two of the strongest plants. Because the plants are such heavy producers, they sometimes exhaust themselves by July or August, so you can start a second crop in early July to keep harvesting into fall.

Like cucumbers, squash plants have separate male and female flowers, and both must be present at the same time for pollination (and fruit production) to occur. Several plants with many flowers will improve the chances of pollination. Provide plenty of water and stand back. Once they start coming, there is abundance.

HARVESTING Check your plants daily. Summer squash should be harvested while they are young and tender, and they can grow several inches overnight. As you harvest, more fruit will come. Mark your calendar for August 8, National Sneak Some Zucchini on Your Neighbor's Porch Day.

VARIETIES Dark green and glossy **'Raven'** (48 to 52 days) is a delicious choice, especially for short-season gardeners. **'Cocozelle'** (53 days) is an Italian heirloom with firm flesh and ridges along its length. Cut slices look like flowers on the plate. The amazing **'Tromboncino'** (70 days) puts overgrown zucchini to shame. These long, twisting squash end in a bulb, which is where all the seeds hide. The entire neck offers up dense, juicy flesh that stays tender, no matter the size. Yellow crookneck **'Horn of Plenty'** (45 to 50 days) lives up to its name, with lots of sweet, firm fruits. **'Sunburst'** (50 to 55 days) is a pattypan, or scallop, squash that glows yellow and has a nutty, earthy taste.

AND HOW ABOUT FRUIT TREES?

Once you realize how delicious your backyard harvest is, you might want to try your hand at growing fruit trees. What's more seasonal than honeyed peaches and crisp apples? Smaller fruit trees make it possible for you to grow and enjoy these sweet treats even if you're gardening in containers on a patio. Go for it.

Apples

Despite the abundance of apples in the Northeast, they are not the easiest fruit to grow. Although New York has great apple-growing conditions, it also has ideal conditions for apple pests and diseases. Rise to the challenge, and you'll enjoy the reward of biting into the sweet crunch of a homegrown fruit.

GROWING Full sun and good drainage are essential; wind protection is a consideration as well. Apples need deep, fertile soil with a pH of 5.5 to 7.0. Spring is the best time to plant apple trees, and young trees are easiest to establish. Mail-order trees are usually 1-year-old whips—trunks with no branches. Nurseries sell potted 2-year-old trees; look for three or four well-spaced branches.

Most apple trees are grafted; plant so that the graft is 2 to 4 inches above ground. If the graft is buried, the top portion of the tree will sprout roots and whatever dwarfing you hoped to gain will be lost. Prune the tree as soon as it is planted: Reduce whips by about a third, just above a bud; cut any branches back by a quarter; and remove crossing branches and those with narrow crotch angles.

Water the trees well and keep them watered throughout summer. Mulch helps keep the developing roots cool and moist. Don't feed the trees their first year; after that you can top dress them annually with a layer of compost or rotted manure. In winter, cover and protect the lower trunks from gnawing rabbits and other critters. The parade of insects that attack apples includes apple maggots, coddling moths, leaf rollers, and aphids. You can foil many apple maggots by hanging several sticky red-ball traps in the tree beginning in mid-June. Coddling moths are trickier; pheromone traps can help. Or you can cover each apple with a small paper bag and staple the ends closed. Applying dormant oil before the flowers open can also help destroy overwintering pests.

Apple scab, fire blight, and cedar apple rust are the most prevalent diseases. Some apple varieties are resistant to apple scab, which overwinters in fallen debris. A wetable sulfur spray can thwart fire blight, but you'll need to apply it several times (check with your local extension office). Cedar apple rust is a disease that jumps back and forth from apple trees to junipers, so choose which tree you want to grow and don't grow the other. Good fall cleanup will also help with all of these problems.

HARVESTING Apple trees start producing in 2 to 7 years. The surest way to judge ripeness is to bite into the fruit. Color is a good indicator, and ripe apples pull easily from the tree, with their stem intact. Late-maturing varieties can be picked a bit immature, but early varieties should be allowed to ripen on the tree.

VARIETIES **'Baldwin'**, with thick, juicy skin and crisp, yellow flesh, is an antique from colonial Massachusetts that stores very well. **'Empire'**, a cross between **'McIntosh'** and **'Red Delicious'**, is a popular eating apple with very sweet, white flesh. Other disease-resistant varieties include **'Freedom'**, **'GoldRush'**, **'Jonafree'**, **'Liberty'**, and **'Redfree'**.

Cherries

I find cherries to be the easiest tree fruit to grow. Sour or tart cherries are used for baking and are self-fertile, so you need only one tree. Sweet cherries usually need a second cultivar for pollination, but they are often grafted with two varieties on one tree, which solves that problem. A few varieties are self-fertile, such as 'Compact Stella'. Cherry trees can grow quite large; consider a dwarf variety for easier harvesting and maintenance.

GROWING Cherry trees need well-draining soil and room to grow and reach their mature size. They bloom early in the spring and are susceptible to frost damage. Plant them out of extreme wind exposure. When the fruit has set, keep the tree evenly watered. Dry weather causes the fruit to shrivel, and too much rain can cause them to crack. Mulching under the trees helps keep the moisture level consistent. Net small trees as soon as the fruits start to show some color to protect them from bird raids. Larger trees should produce enough to share. Prune only to allow light into the center of the tree. Sweet cherries can be pruned on top every 3 or 4 years to promote more lower branching.

HARVESTING When you notice a few cherries lying on the ground, it's time to harvest. Gently pull the clusters from the branches, keeping the stems on the cherries.

VARIETIES *Sweet*: **'Rainier'** (zones 4 to 8) is a large tree with firm fleshed yellow and red fruits. **'Stella'** (zones 5 to 8) cherries are extremely sweet and resist cracking. *Sour*: **'Montmorency'** (zones 4 to 9) is the classic pie cherry. **'English Morello'** (zones 4 to 9) ripens later in spring and can be sweet enough to eat fresh.

Peaches and Nectarines

Peach and nectarine trees are pretty easy to maintain, as fruit trees go. They do require pruning, but a little effort once a year is not much to ask. I strongly recommend a dwarf tree, because a height of 8 to 10 feet is easier to maintain and harvest than a tree that grows to 15 feet or taller.

GROWING Late spring frosts are the biggest enemy; a sudden dip to freezing can kill an entire season's buds. Don't plant in a windy area or at the bottom of a hill, where frost tends to settle and linger. Peaches and nectarines need full sun, a well-drained loamy soil, and a neutral pH around 6.5. Neither peaches nor nectarines need a cultivar for cross pollination, so you need only one tree. It's common to find good sized, potted peach trees for sale at garden centers, but if you mail-order your tree, you will probably get a 1-year-old whip, which is a 3- to 4-foot-tall trunk with no branches.

Peaches and nectarines are pruned or trained using the open center system—basically a vase shape with two or three major (scaffold) limbs. This configuration allow lots of air and sun into the center of the tree. You'll need to prune every year to maintain this form. Fruit appears on the prior year's growth, so pruning each year ensures new growth. Prune a 1-year-old tree immediately after planting to a height of 26 to 30 inches and remove all the branches. It seems severe, but this allows you to shape the tree from the start as new branches appear. As the tree grows, remove any low-hanging limbs and any upright shoots in the center of the tree. In year 2, choose two or three strong, wide-angled limbs as scaffolds and cut off all the others. Prune the scaffolds to 30 inches. Then in subsequent years, remove anything growing up or into the center of the tree. You can maintain the height of the tree by pruning scaffold branches back to 8 to 10 feet.

Mature trees will set more fruit than they can handle. Some fruits will be shed naturally in early summer, but this may not be enough, so you'll need to remove some more by hand to ensure a large and healthy crop. When the fruits are about the size of a quarter, thin them to a spacing of 6 to 8 inches to allow the remaining fruits to grow into large, sweet peaches or nectarines. To keep your tree growing and setting new buds every year, feed annually in early spring with a fruit tree fertilizer or any balanced fertilizer: 1 pound for young trees and 2 pounds for mature bearing trees.

Watch out for the tarnished plant bug, peach tree borers, and the plum curculio. The most common problem is the peach leaf curl disease.

VARIETIES *Peaches*: **'Contender'** (zones 4 to 8) withstands late-season frosts. Yellow-fleshed **'Redhaven'** (zones 5 to 8) is considered the standard for peach quality, and it doesn't drop its fruit. **'Reliance'** (zones 4 to 8) has honey-sweet fruits and is considered one of the most winter-hardy peaches. *Nectarines*: **'Mericrest'** (zones 5 to 8) has excellent flavor and good disease resistance. **'Hardired'** (zones 5 to 8) tolerates bacterial spot and brown rot.

RESOURCES

SEED AND PLANT CATALOGS

American Meadows Inc.
Williston, Vermont
americanmeadows.com
What started as a source for wildflower seeds has expanded into an online garden community with resources that include all kinds of seed, plants, supplies and plenty of growing information.

Baker Creek Heirloom Seeds
Mansfield, Missouri
rareseeds.com
Baker Creek offers a huge selection of vegetable seeds from around the world, all open-pollinated and many heirloom.

Berlin Seeds
Millersburg, Ohio
330-893-2091
Based in Amish country, Berlin Seeds currently sells only by phone and in their store. The catalog offers a good cross section of seeds, lots of unusual tools and excellent growing advice.

Fedco Seeds
Waterville, Maine
fedcoseeds.com
All seeds are tested and proven for even the coldest Northeast garden.

Gourmet Seed International, LLC
Tatum, New Mexico
gourmetseed.com
Uncommon and specialty varieties are the attraction here. This family-run business is focused on high quality, with hermetically-sealed, long-life packaging and an international selection of seed.

Heirloom Seeds
West Finley, Pennsylvania
heirloomseeds.com
This is another small family-run operation with a large selection of heirloom and open pollinated varieties. If you can't decide what to try, you can choose a pre-packaged complete garden kit.

High Mowing Organic Seeds
Wolcott, Vermont
highmowingseeds.com
You'll find hundreds of vegetable, herb, and flower seed varieties, plus a nice selection of cover crops.

Hudson Valley Seed Company
Accord, New York
hudsonvalleyseed.com
This company started as a seed lending library. The catalog focuses on Northeast gardens. All seed is hand grown, harvested, and packed.

John Scheepers Kitchen Garden Seeds
Bantam, Connecticut
kitchengardenseeds.com
The sister company of Van Engelen Bulbs, Scheepers offers all kinds of vegetables and herbs, as well as recipes for using them.

Johnny's Selected Seeds
Waterville, Maine
johnnyseeds.com
This employee-owned company specializes in vegetable varieties bred for flavor, ease of growing, and short seasons.

J.W. Jung Seed Company
Randolph, Wisconsin
jungseed.com
A family run business for over 100 years. You'll find a little bit of everything in their iconic catalog, from All America Selections to heirloom seeds and plants.

NE Seed
Hartford, Connecticut
neseed.com
Offering a mix of heirlooms and tested hybrids, NE Seed supplies bulk seed to small growers and packets to backyard gardeners. You can even find Wildlife Food Plot Mixes.

Park Seed
Greenwood, South Carolina
parkseed.com
Park Seed is one of the oldest seed and mail-order companies in the United States, selling a vast listing of edible and ornamental seed, plants, trees, and shrubs.

Peaceful Valley Farm & Garden Supply
Grass Valley, California
groworganic.com
Peaceful Valley is one of the largest suppliers of organic growing supplies. Their extensive seed listings include a nice assortment of cover crops and bulk offerings.

Pinetree Garden Seeds
New Gloucester, Maine
superseeds.com
Pinetree caters to home gardeners, with a large selection of seeds and plants offered in smaller quantities (and prices).

Raintree Nursery
Morton, Washington
raintreenursery.com
Although they're located on the other side of the continent, Raintree is an excellent source of fruit trees and berry plants.

Sand Hill Preservation Center
Calamus, Iowa
sandhillpreservation.com
This family-run farm and seed business specializes in open-pollinated and heirloom varieties. They're a great source for sweet potato slips, and they also sell poultry.

Seed Savers Exchange
Decorah, Iowa
seedsavers.org
Seed Savers is a renowned non-profit seed bank and a terrific source of heirloom and open-pollinated seeds. Members have access to hundreds of varieties made available by other members.

Seeds from Italy
Lawrence, Kansas
growitalian.com
This source for European heirlooms is the North American arm of Franchi Seed, Italy's oldest family-owned seed company.

Southern Exposure Seed Exchange
Mineral, Virginia
southernexposure.com
Southern Exposure's emphasis is on varieties for the mid-Atlantic and U.S. Southeast. They offer more than 700 varieties of vegetable, flower, herb, grain, and cover crop seeds.

Stark Bro's
Louisiana, Missouri
starkbros.com
This company has been selling fruit, nut, and berry plants to backyard gardeners for almost 200 years.

Tomato Growers Supply Company
Fort Myers, Florida
tomatogrowers.com
If you love tomatoes, peppers and eggplants, prepare to spend some time perusing this catalog. You'll find varieties divided and listed by harvest times, to keep your garden growing all season.

Totally Tomatoes
Randolph, Wisconsin
totallytomato.com
What started as a source for tomato seeds has grown into a company that sells all kinds of seeds, but they still wow with their selection of tomatoes.

Turtle Tree Seed
Copake, New York
turtletreeseed.org
This small non-profit company sells 100-percent open-pollinated, organic vegetable, herb, and flower seeds.

White Harvest Seed Company
Hartville, Missouri
whiteharvestseed.com
A relatively new family-run business that specializes in heirloom seeds. They'll even help you layout your garden and plan how much to plant in your garden.

Wood Prairie
Bridgewater, Maine
woodprairie.com
The specialty of this small, Certified Organic family farm is seed potatoes, along with potato growing and cooking info and a Potato Helpline. Other vegetable seeds are now also available.

REGIONAL RESOURCES

Backyard Fruit Growers
byfg.org
This informal association, based in Lancaster, Pennsylvania, provides information on growing tree fruits and berries. Members receive a newsletter and can attend tours, fruit-tastings, and workshops.

Northeast Organic Farming Association (NOFA)
nofa.org
With chapters throughout the Northeast, NOFA offers growing information, conferences, and workshops to farmers, gardeners, and anyone who works with the land.

COOPERATIVE EXTENSION SERVICES HOME GARDENING

gardening.cals.cornell.edu/garden-guidance/foodgarden
Their fact sheets offer information about growing, caring for, and harvesting crops in a backyard garden. Contact your local cooperative extension office and Master Gardener program from its web site. (Note that web site addresses may change.)

NON-PROFIT ORGANIZATIONS

Ample Harvest
ampleharvest.org
This national program encourages gardeners to donate their surplus harvest to local food pantries.

Plant a Row for the Hungry
www.gardenwriters.org/PAR
This program encourages gardeners to plant an extra row of vegetables and donate their surplus to local food agencies and soup kitchens. Since the program began in 1995, more than 2 million pounds of produce per year have been donated by American gardeners.

WEATHER SERVICES AND ZONE MAPS

National Oceanic and Atmospheric Administration, U.S. Climate Normals
climate.gov/tags/us-climate-normals
NOAA's National Climatic Data Center offers state-by-state averages for the last three decades of first and last frost dates, growing degree dates, and other statistical information.

USDA plant hardiness zone map
planthardiness.ars.usda.gov

SOIL TESTING

You can have soil tested through most cooperative extension agencies or the following laboratories.

Cornell Nutrient Analysis Laboratory
Ithaca, New York
cnal.cals.cornell.edu/

Soil & Plant Testing Laboratory
Amherst, Massachusetts
http://ag.umass.edu/services/soil-plant-nutrient-testing-laboratory

FURTHER READING

The more I garden, the more questions I have, so I always keep several books handy. My favorite gardening books offer a mix of education, encouragement, and inspiration. This is a list of some that I turn to so often, they feel like old friends and garden mentors. It's always nice to compare notes with someone who shares your devotion for gardening, and I hope this book will become a tried-and-true reference for you, both during the growing season and to get you through the seemingly endless days of winter.

Ashworth, Suzanne. 2002. *Seed to Seed: Seed Saving and Growing Techniques for the Vegetable Gardener*. White River Junction, Vermont: Chelsea Green Publishing.

Coleman, Eliot. 1999. *Four Season Harvest: Organic Vegetables from Your Home Garden All Year Long*. White River Junction, Vermont: Chelsea Green Publishing.

Cunningham, Sally Jean. 2000. *Great Garden Companions: A Companion-Planting System for a Beautiful, Chemical-Free Vegetable Garden*. Emmaus, Pennsylvania: Rodale Press.

Deardorff, David, and Kathryn Wadsworth. 2011. *What's Wrong With My Vegetable Garden? 100% Organic Solutions for All Your Vegetables, from Artichokes to Zucchini*. Portland, Oregon: Timber Press.

Eierman, Colby. 2012. *Fruit Trees in Small Spaces: Abundant Harvests from Your Own Backyard*. Portland, Oregon: Timber Press.

Jabbour, Niki. 2011. *The Year-Round Vegetable Gardener: How to Grow Your Own Food 365 Days a Year, No Matter Where You Live*. North Adams, Massachusetts: Storey Publishing.

Nichols McGee, Rose Marie, and Maggie Stuckey. 2002. *Bountiful Container: Create Container Gardens of Vegetables, Herbs, Fruits, and Edible Flowers*. New York: Workman Publishing.

Reich, Lee. 1993. *A Northeast Gardener's Year*. New York: Da Capo Press.

———. 2012. *Grow Fruit Naturally: A Hands-On Guide to Luscious, Home-Grown Fruit*. Newtown, Connecticut: Taunton Press.

Riotte, Louise. 1998. *Carrots Love Tomatoes: Secrets of Companion Planting for Successful Gardening*. North Adams, Massachusetts: Storey Publishing.

PLANT HARDINESS ZONES

ZONE	TEMPERATURE (°F)
1	Below –50
2	–50 to –40
3	–40 to –30
4	–30 to –20
5	–20 to –10
6	–10 to 0
7	0 to 10
8	10 to 20
9	20 to 30
10	30 to 40
11	40 to 50
12	50 to 60
13	60 to 70

PHOTOGRAPHY AND ILLUSTRATION CREDITS

PHOTOGRAPHY

Alamy
Mint Images Limited: 10, 16; Tony Watson: 140

Jackie Connelly: 132

GAP Photos
Gary Smith: 48

Marie Iannotti: 180, 222, 225

iStockphoto
encrier: 2; Boogich: 15; Atom Studios: 18; space-monkey-pics: 32; BruceBlock: 35; sanddebeautheil: 46–47; IPGGutenbergUKLtd: 64; alle12: 127; sagarmanis: 146, 181; Pavliha: 162–163; Gooddenka: 182; Quanthem: 185; Eugenegg: 189; efilippou: 190; KChodorowski: 191; TasiPas: 196; duckycards: 210; johnnyscriv: 213; moisseyev: 214; Milkos: 215; jlmcloughlin: 216; Yoeml: 217; Mindstyle: 218; smartstock: 220; Danler: 224; zeleno: 226; Cellena: 229

Pixabay
Jenny Johannsson: 176; 858106: 179; fraban: 195; evitaochel: 221; BabaMu: 223

PxHere
Used under a Creative Commons 1.0 License Public Domain Dedication: 8–9, 56, 74, 98, 183, 188, 193, 200, 203, 204, 205, 206, 207, 211, 212, 219, 227, 228

Shutterstock
Fabio Pagani: 21; Rainier Fuhrman: 23; Nadzeya Pakhomav: 27; Boumen Japet: 31; Del Boy: 38; jkcDesign: 41; Rawpixel.com: 43; Michelle Lee Photography: 51; Skeronov: 53; vitec: 69; Phil Darby: 71; Jetanoom: 79; Kosobu: 83; Ulza: 88; Milante: 93; Eric Krouse: 95; Alexander Raths: 105, 110; Miyuki Satake: 116; Grandiflora: 122; mcajan: 130; mnimage: 138; PFlemingWeb: 151; AForlenza: 156; vaivirga: 177; Bastiaanimage stock: 192; Svekrova Olga: 194; Denis Pogostin: 199; pixfix: 201

Wikimedia Commons
Used under a Creative Commons Attribution–Share Alike 2.0 Generic license
H. Alexander Talbot: 198

Used under a Creative Commons Attribution–Share Alike 3.0 Unported license
Earth 100: 178, 208; Jamain: 187, 197

Released into the Public Domain
Maria Chantal Rodriguez Nilsson: 137; Stephen Ausmus, USDA ARS: 209

ILLUSTRATIONS

David Deis/Dreamline Cartography: 13

iStockphoto
solargaria: 5; aqua_marinka: 47, 106; kateja_f: 62, 63

Julia Sadler: 24, 29, 39, 40, 42, 54, 55, 59, 61, 72, 80, 97, 103, 109, 115, 131, 152

Shutterstock
maritime_m: 9, 73, 164; sdp_creations: 54; MorePics: 107; ArtMari: 120; Ann Doronina: 185

INDEX

Main entry pages for Edibles A to Z appear in bold type.

C

D

E

F

R

T

U

V

W

Z

JULIET LOFARO

Marie Iannotti is a gardener who writes, photographs, and speaks lovingly and irreverently about gardening. She has been an avid grower and eater of vegetables since her parents started her off as chief bean picker in her family garden. Her first book, *The Beginner's Guide to Heirloom Vegetables*, grew out of her quest to find the best-tasting vegetable varieties. Marie is a Master Gardener emeritus, as well as a former Cornell Cooperative Extension Horticulture Educator. She was the gardening expert at About.com for over a decade. To date, she has written three books and her writing has been featured in newspapers and magazines nationwide. You can find her writing about gardening for the joy of it at MarieI.com.